Lecture Notes in Mathematics

Edited by A. Dold and B. Eckmann

T0216161

580

C. Castaing
M. Valadier

Convex Analysis and Measurable Multifunctions

Springer-Verlag
Berlin · Heidelberg · New York 1977

Authors
Charles Castaing
Michel Valadier
Université des Sciences
et Techniques du Languedoc
Place Eugène Bataillon
34060 Montpellier Cedex/France

Library of Congress Cataloging in Publication Data

Castaing, Charles, 1932-
Convex analysis and measurable multifunctions.

(Lecture notes in mathematics ; 580)
Includes bibliographies and index.
1. Functional analysis. 2. Convex functions.
I. Valadier, M., 1940- joint author. II. Title.
III. Series: Lecture notes in mathematics (Berlin) ; 580
QA3.L28 no. 580 [QA320] 510'.8s [515'.7] 77-3987

AMS Subject Classifications (1970): 46XX

ISBN 3-540-08144-5 Springer-Verlag Berlin · Heidelberg · New York
ISBN 0-387-08144-5 Springer-Verlag New York · Heidelberg · Berlin

2141/3140-543210

Preface

The present work is devoted to convex analysis, measurable multi-functions and some of their applications. The only necessary prerequisite for an intelligent reading is a good knowledge of analysis (Bourbaki or Dunford-Schwartz are appropriate references). One exception is the use of liftings of L^∞; for their existence we refer to Ionescu-Tulcea's book.

Many questions are not treated, for example: the Borel selection theorem due to Novikov, Arsenin, Kunugui ... ; the theory of set valued measures (Artstein, Costé, Drewnowsky, Godet-Thobie, Pallu de La Barrière ...); the set valued martingales (Bismut, Daurès, Neveu, Van Cutsem ...); the application to optimal control and to the calculus of variations (Ekeland-Temam, Olech, Rockafellar ...).

Each chapter has its own bibliography. Apologies are offered in advance to those who feel that they have been slighted.

We take this opportunity to thank a small group of colleagues for their help in revising our manuscript. Finally, thanks are due to Mme Mori who typed most of the text, to M. Meyran and the whole secretariat of the department of mathematics.

Montpellier, October 1975

Contents

Chapter I

CONVEX FUNCTIONS

Measurable convex valued multifunctions (and more generally convex integrands) have acquired great importance in recent years, and will be treated in later chapters.

We intend in this chapter to give briefly some basic results on convex functions, which cannot be found in Bourbaki or Dunford Schwartz. That is the modern theory built up by Fenchel, Moreau, Rockafellar and some others (see historical comments and bibliography in Moreau [11]).

The possibility for a convex function to take the value $+\infty$ permits us to consider only functions which are defined on a whole linear space (and not on a subset). The polar of a function generalizes the notion of a support function of a set, and is the basis of duality arguments. The main results of the theory are those about differentiability (theorems 27, 28, 29), which apply particularly to optimisation problems, and also inf-compactness properties, which furnish existence theorems. Infimum-convolution has not so direct applications, but is closely related to the other notions (because roughly speaking inf-convolution and addition are mutually polar operations), and it can be considered as one of the geometrical sides of the theory.

§ 1 - CONVEX LOWER SEMI-CONTINUOUS FUNCTIONS. BIPOLAR THEOREM -

1 - We recall some definitions and basic facts about functions. Let E be a real topological vector space and $f : E \to \bar{\mathbb{R}} = [-\infty, \infty]$.

The <u>effective domain</u> of f is

$$\operatorname{dom} f = \{x \in E \mid f(x) < \infty\}.$$

The <u>epigraph</u> of f is

$$\operatorname{epi} f = \{(x, r) \in E \times \mathbb{R} \mid r \geq f(x)\}.$$

The function f is <u>convex</u> if for every $x, y \in E$, $\lambda \in [0, 1]$, $f(\lambda x + (1 - \lambda)y) \leq \lambda f(x) \dot{+} (1 - \lambda) f(y)$ (with the convention $(+\infty) \dot{+} (-\infty) = +\infty$). It is equivalent to suppose, dom f is convex and the restriction of f to dom f (with values in $[-\infty, \infty[$) is convex. It is also equivalent to suppose epi f is convex in $E \times \mathbb{R}$. Convex functions taking value $-\infty$ are very special, and will be often avoided.

If C is a convex subset of E and $f : C \to \bar{\mathbb{R}}$ is convex, then the function $\bar{f}$ defined by

$$\bar{f}(x) = \begin{cases} f(x) & \text{if } x \in C \\ +\infty & \text{if } x \in E - C \end{cases}$$

is convex, and this extension is very convenient.

The function f is <u>lower semi-continuous</u> if and only if epi f is closed. For any function f there exists a greatest l.s.c. function $\bar{f}$ less than f, whose epigraph is $\operatorname{epi} \bar{f} = \overline{\operatorname{epi} f}$. Furthermore $\bar{f}(x) = \varliminf_{y \to x} f(y)$.

For any function f there exists a greatest convex l.s.c. function less than f, denoted $\overline{\text{co}}\ f$, whose epigraph is $\operatorname{epi}(\overline{\text{co}}\ f) = \overline{\text{co}}(\operatorname{epi} f)$ (in the right member $\overline{\text{co}}$ denotes the closed convex hull).

2 - Définition - Let E be a topological vector space and $f : E \to \overline{\mathbb{R}}$. The polar function of f is the function $f^* : E' \to \overline{\mathbb{R}}$ defined by $f^*(x') = \sup \{\langle x', x\rangle - f(x) \mid x \in E\}$.

For symmetry it could be more convenient to consider two vector spaces E and F in separate duality. Thus F is the dual of E with the weak topology $\sigma(E, F)$ or other topologies compatible with duality.

For examples see § 6.

An important and obvious property is that $f \le g$ implies $f^* \ge g^*$.

When f is the indicator function of a set A, that is

$$f(x) = \begin{cases} 0 & \text{if } x \in A \\ \infty & \text{if } x \notin A \end{cases}, \quad \text{denoted } \delta(x|A),$$

the polar function is given by

$$\delta^*(x'|A) = \sup \{\langle x', x\rangle \mid x \in A\}$$

and so is the support function of A.

Remark. There is a relation between the polar function of f and the support function of epi f. Suppose f is not the constant function $+\infty$ (that is epi $f \neq \emptyset$). Then

$$\delta^*((x', r)|\text{epi } f) = \begin{cases} +\infty & \text{if } r > 0 \\ \delta^*(x'|\text{dom } f) & \text{if } r = 0 \\ -r\, f^*(-\frac{x'}{r}) & \text{if } r < 0. \end{cases}$$

In particular $f^*(x') = \delta^*((x', -1)|\text{epi } f)$.

If $f(x) = +\infty$ for every x, $f^*(x') = -\infty$ and $\delta^*((x', r)|\text{epi } f) = -\infty$.

3 - Theorem I.3 - Let E be a Hausdorff locally convex space and $f : E \to]-\infty, \infty]$ be convex and l.s.c. Then f is the supremum of all affine continuous functions less than f.

Proof. 1) Let us prove that there exists a continuous affine function less than f. The case $f = +\infty$ is obvious. Suppose there exists x_o such that $f(x_o) < \infty$. Let $r_o \in \mathbb{R}$, $r_o < f(x_o)$. Then epi f is a closed convex set disjoint of $\{(x_o, r_o)\}$. By separation theorem (see for example Bourbaki II-5-3 Prop. 4, p. 84) there exists $(x'_o, \lambda_o) \in E' \times \mathbb{R}$ which separates strictly epi f and $\{(x_o, r_o)\}$. We may suppose

$$(1)\quad \begin{cases} \text{for every } (x, r) \in \text{epi } f \\ \langle (x'_o, \lambda_o), (x, r) \rangle \geq \alpha > \langle (x'_o, \lambda_o), (x_o, r_o) \rangle. \end{cases}$$

For $(x, r) = (x_o, f(x_o))$ (1) gives $\lambda_o f(x_o) > \lambda_o r_o$, hence $\lambda_o > 0$. For any x one has

$$\langle x'_o, x \rangle + \lambda_o f(x) \geq \alpha, \quad \text{hence}$$

$$f(x) \geq \frac{1}{\lambda_o} [-\langle x'_o, x \rangle + \alpha].$$

Let $p(x) = \frac{1}{\lambda_o} [-\langle x'_o, x \rangle + \alpha]$. Then p is a continuous affine function less than f.

2) The theorem will be proved if we prove that for $x_1 \in E$ and $r_1 < f(x_1)$ there exists a continuous affine function q less than f, such that $q(x_1) > r_1$. Again by the separation theorem there exists (x', λ) and $\beta \in \mathbb{R}$ such that

$$(2)\quad \begin{cases} \text{for every } (x, r) \in \text{epi } f \\ \langle (x', \lambda), (x, r) \rangle \geq \beta > \langle (x', \lambda), (x_1, r_1) \rangle \end{cases}$$

As epi $f \supset \{x_o\} \times [f(x_o), \infty[$, $\lambda \geq 0$. If $\lambda > 0$, one has, from

$$\langle x', x \rangle + \lambda f(x) \geq \beta,$$

$$f(x) \geq \frac{1}{\lambda} [-\langle x', x \rangle + \beta].$$

If $\lambda = 0$, one has

$$< x', x > \geq \beta > < x', x_1 > \text{ for every } x \in \text{dom } f.$$

Consider the affine continuous function (p is defined in the first part of the proof)

$$q(x) = p(x) + k\ [\beta - < x', x >] \text{ for } k \geq 0.$$

Then $q \leq f$, and for k large enough one has $q(x_1) > r_1$.

Remark The set of all functions $f : E \to \overline{\mathbb{R}}$ which are supremum of a family of continuous affine functions is **denoted by** $\Gamma(E)$. Theorem 3 describes these functions except the constant $-\infty$ which is the supremum of the empty family. A convex l.s.c. function **is said to be** proper if it is not the constant $+\infty$ and if it does not take the value $-\infty$. The set of all these functions is **denoted by** $\Gamma_0(E)$. Then $\Gamma(E) = \Gamma_0(E) \cup \{-\infty\} \cup \{+\infty\}$ (here $\pm\infty$ denotes the corresponding constant function on E).

4 - Theorem I.4 - Let E be a Hausdorff locally convex space, E' its dual. Let $f : E \to \overline{\mathbb{R}}$, f^* its polar and f^{**} its bipolar :

$$f^{**}(x) = \sup \{< x', x > - f^*(x') \mid x' \in E'\}.$$

Then f^{**} is the supremum of all continuous affine functions less than f. Moreover if $f \in \Gamma(E)$, $f = f^{**}$, and $f \mapsto f^*$ is a one to one map from $\Gamma(E)$ to $\Gamma(E')$ (or from $\Gamma_0(E)$ to $\Gamma_0(E')$) with inverse map $f^* \mapsto f^{**}$.

Proof. If $f = +\infty$, $f^* = -\infty$ and $f^{**} = +\infty$. Suppose now that f is not the constant $+\infty$, so that $f^*(x') > -\infty$ for every x'. Let $x' \in E'$, and $\alpha \in \mathbb{R}$. The continuous affine function $< x', . > - \alpha$ is less than f if

$< x', x > - \alpha \leq f(x)$ for every $x \in E$. That is equivalent to $\alpha \geq \sup \{< x', x > - f(x) | x \in E\} = f^*(x')$.

Denote by $A(x')$ the set of all continuous affine functions $< x', . > - \alpha$ $(\alpha \in \mathbb{R})$ less than f. Then if $f^*(x') \in \mathbb{R}$, $A(x')$ has a greatest member which is $< x', x > - f^*(x')$.
If $f^*(x') = \infty$, $A(x')$ is empty. Now it is clear that $\sup \{< x', . > - f^*(x') | x' \in E\}$ is the supremum of all continuous affine functions less than f.

The last part of the theorem follows from n° 3.

5 - Theorem I.5 - Let E be a Hausdorff locally convex space and $f : E \to]-\infty, \infty]$. Suppose there exists at least one continuous affine function less than f. Then the greatest convex l.s.c. function less than f is the bipolar f^{**}.

Proof. Remark first that the greatest convex l.s.c. function less than f, $\overline{co}\, f$, verifies

$\overline{co}\, f \geq f^{**}$. But $\overline{co}\, f$ belongs to $\Gamma(E)$ because it is greater than a continuous affine function. Then

$$(\overline{co}\, f)^{**} = \overline{co}\, f .$$

From $f \geq \overline{co}\, f$ follows $f^{**} \geq (\overline{co}\, f)^{**}$.

Thus $\overline{co}\, f = f^{**}$.

Remark. A convex l.s.c. function f which takes the value $-\infty$ has the following form :

$$\begin{cases} \text{dom } f \text{ is a closed convex non empty set} \\ f(x) = -\infty \quad \text{if} \quad x \in \text{dom } f. \end{cases}$$

For such a function $f = \overline{co}\, f$, but $f^{**} = -\infty$.

§ 2 - SOME PROPERTIES OF CONVEX SETS -

6 - Let E be a vector space and $A \subset E$. Then the gauge of A is the function $j(x) = \inf \{k \in]0, \infty[\mid x \in kA\}$ (the infimum is taken in $[0, \infty]$ or $\overline{\mathbb{R}}$ so that $\inf \emptyset = +\infty$).

If for every line $\mathbb{R}x$ $(x \neq 0)$, $A \cap \mathbb{R}x$ is closed and convex and contains 0, then :

- $j(x) = \infty$ if $A \cap \mathbb{R}_+ x = \{0\}$
- $j(x) = 0$ if $A \cap \mathbb{R}_+ x = \mathbb{R}_+ x$
- and if $A \cap \mathbb{R}_+ x = [0, \lambda x]$ with $\lambda > 0$, $j(x) = \frac{1}{\lambda}$

In a normed space the gauge of the unit ball is the given norm.

Lemma I-6. If A is a closed convex set containing 0, then the gauge of A, j, is convex l.s.c., positively homogeneous,

$$A = \{x \in E \mid j(x) \leq 1\},$$

and

$$j^* (x') = \delta(x' \mid A^0)$$

Moreover epi j is the closed cone generated by $A \times \{1\}$.

Proof. Let $x, y \in E$ and $\alpha \in [0, 1]$. We want to prove

$$j(\alpha x + (1 - \alpha)y) \leq \alpha j(x) + (1 - \alpha) \ j(y).$$

We may suppose $j(x) < \infty$ and $j(y) < \infty$. Then if $\varepsilon > 0$

$$x \in (j(x) + \varepsilon) A \text{ and } y \in (j(y) + \varepsilon)A. \text{ Hence}$$

$$\alpha x + (1 - \alpha) y \in \alpha(j(x) + \varepsilon)A + (1 - \alpha)(j(y) + \varepsilon)A$$

$$= [\alpha j(x) + (1 - \alpha) j(y) + \varepsilon]A$$

That entails

$$j(\alpha x + (1 - \alpha) y) \leq \alpha j(x) + (1 - \alpha) j(x). \text{ Thus } j \text{ is convex.}$$

It is clear that j is positively homogeneous, and that

$A = \{x \in E \mid j(x) \leq 1\}$. Therefore for every $\alpha \in \mathbb{R}$

$$\{x \in E \mid j(x) \leq \alpha\} = \begin{cases} \emptyset \text{ if } \alpha < 0 \\ \alpha A \text{ if } \alpha > 0 \\ \bigcap_{\beta > 0} \beta A \text{ if } \alpha = 0 \end{cases}$$

and j is l.s.c.

By the property of j on every half line $\mathbb{R}_+ x$ $(x \neq 0)$ it is clear that epi j is the closed cone generated by $A \times \{1\}$.

It remains to prove

$j(x) = \delta^*(x|A^\circ)$. That results from the fact that for every $\alpha > 0$

$\delta^*(x|A^\circ) \leq \alpha \Leftrightarrow x \in \alpha A^{\circ\circ} \Leftrightarrow x \in \alpha A$.

7 - The following proposition describes the asymptotic cone of a closed convex set.

Proposition I-7. *Let* E *be a Hausdorff topological vector space and* C *be a closed convex non empty set. For every* $x_0 \in C$ *there exists a greatest cone* Λ *such that* $x_0 + \Lambda \subset C$: $\Lambda = \bigcap_{\lambda > 0} \lambda(C - x_0)$. *This cone does not depend on* x_0, *is closed convex and is called the asymptotic cone of* C. *Moreover if* E *is locally convex* $\Lambda = [\mathrm{dom}(\delta^*(.|C))]^\circ$. *This cone is denoted* $\mathrm{As}(C)$.

Proof. It is obvious that the greatest cone Λ such that $x_0 + \Lambda \subset C$ is $\bigcap_{\lambda > 0} \lambda(C - x_0)$.

It is easy to see that for $y \in E$, $x_0 \in C$ and $x_1 \in C$, $x_0 + \mathbb{R}_+ y \subset C \Leftrightarrow x_1 + \mathbb{R}_+ y \subset C$ (because C is closed and convex). Hence Λ does not depend on x_0.

Finally, if E is locally convex, if $x_0 \in C$ and $y \in E$,

$y \in \Lambda \Leftrightarrow \forall r \geq 0,\ x_0 + ry \in C$

$\Leftrightarrow \forall r \geq 0,\ \forall x' \in \mathrm{dom}\ \delta^*(.|C),\ \langle x', x_0 + ry \rangle \leq \delta^*(x'|C)$

$\Leftrightarrow \forall x' \in \mathrm{dom}\ \delta^*(.|C),\ \langle x', y \rangle \leq 0.$

8 - In the following lemma we use generalized sequences and ultrafilters. It is possible to use only filters (see Dieudonné [5]). But to prove theorem 10 below, it is very natural to use generalized sequences (because if the space is metrizable, the most simple proof uses ordinary sequences). And to obtain a cluster point under some compactness hypotheses, ultrafilters are very convenient.

We recall that a generalized sequence $(x_i)_{i \in I}$ is a family whose index set is a directed ordered set. The set of all sections $\{\{j \in I \mid j \geq i\} \mid i \in I\}$ is a basis of filter on I. When the points x_i belong to a topological space, one can define in an obvious way the limit, or a cluster point, of the generalized sequence.

Lemma I-8. Let E be a Hausdorff topological vector space, C a closed convex locally compact set, $(x_i)_{i \in I}$ a generalized sequence in C, and $\mathcal{U}$ an ultrafilter on I finer than the filter of sections. Suppose $0 \in C$ and let V be a closed circled (circled means that for every $r \in [-1, 1]$, $rV \subset V$) neighbourhood of 0 such that $C \cap V$ is compact. Denote by j the gauge of V. Then either:

$$\lim_{\mathcal{U}} j(x_i) < \infty \text{ and } (x_i) \text{ converges}$$

or $\lim_{\mathcal{U}} j(x_i) = \infty$ and $\dfrac{x_i}{j(x_i)}$ converges to a non null vector of $As(C)$.

Proof. If $\lim j(x_i) < \infty$ there exists $n \in \mathbb{N}$ and $J \in \mathcal{U}$ such that $j(x_i) \leq n$ for every $i \in J$. But $C \cap nV \subset n(C \cap V)$, so $C \cap nV$ is compact. Hence (x_i) converges.

If $\lim j(x_i) = \infty$, for every n there exists $J_n \in \mathcal{U}$ such that $j(x_i) \geq n$ for every $i \in J_n$. Then $j(x_i) \neq 0$ for every i in J_1.

For such i, $\dfrac{x_i}{j(x_i)} \in V - \frac{1}{2}\mathring{V}$. So $\dfrac{x_i}{j(x_i)}$ converges in $C \cap (V - \frac{1}{2}\mathring{V})$ to a non null vector y. It remains to prove $y \in As(C)$.

For any integer $n \geq 1$ and every $i \in J_n$, $\frac{x_i}{j(x_i)} \in \frac{1}{n} C$. Hence $y \in \frac{1}{n} C$. Therefore $y \in \bigcap_{n \geq 1} \frac{1}{n} C = As(C)$.

9 - Theorem I-9. *Let E be a Hausdorff topological vector space and C a closed convex locally compact non empty set. If $As(C) = \{0\}$, C is compact.*

Proof. By lemma 8 every generalized sequence $(x_i)_{i \in I}$ in C has a cluster point because the second case in lemma 8 cannot occur. Hence C is compact.

Remark. Another proof is the following : if C is not compact, then (with notations of the lemma) C is not contained in a set nV. Put $x_n \in C - nV$. Then by lemma 8 applied to $(x_n)_{n \in \mathbb{N}}$, the asymptotic cone is not $\{0\}$.

10 - Theorem I-10. *Let E be a Hausdorff topological vector space, and C and D two closed convex subsets of E. Suppose C is locally compact and $As(C) \cap (-As(D)) = \{0\}$. Then C + D is closed.*

Proof. We may suppose C and D non empty, and $0 \in C \cap D$. Let $x \in \overline{C + D}$. Then $x = \lim (c_i + d_i)$, where $(c_i)_{i \in I}$ and $(d_i)_{i \in I}$ are generalized sequences in C and D with the same index set. (I may be a basis of neighbourhoods of 0). Let $\mathfrak{U}$ be an ultrafilter on I finer than the filter of sections. Then lemma 8 applies to C, $(c_i)_{i \in I}$ and $\mathfrak{U}$. If (c_i) converges, so does (d_i) and $x = \lim_{\mathfrak{U}} x_i + \lim_{\mathfrak{U}} d_i \in C + D$. Consider now the second case of lemma 8. Then $\frac{c_i}{j(c_i)} \to \bar{y} \in As(C)$ and $\bar{y} \neq 0$. As $j(c_i) \to \infty$, $j(c_i) \neq 0$ for every i in a suitable $J \in \mathfrak{U}$. For such i, as $\frac{x}{j(c_i)} \to 0$, we have $\frac{d_i}{j(c_i)} \to -\bar{y}$.

For every $n \geq 1$, there exists $J \in \mathfrak{U}$ such that for every $i \in J$,

$j(c_i) \geq n$. Hence $\frac{d_i}{j(c_i)} \in \frac{1}{n} D$, $-\bar{y} \in \frac{1}{n} D$ and $-\bar{y} \in \bigcap_{n \geq 1} \frac{1}{n} D = As(D$

That contradicts the hypothesis $As(C) \cap (-As(D)) = \{0\}$.

Remark. For another result see Choquet [4].

- Definition. Let E be a Hausdorff locally convex space and C be a cone in E. Suppose there exists a closed hyperplane H such that $0 \notin H$ and that C is the cone generated by $C \cap H$. Then the set $C \cap H$ is called a base of C.

Lemma I-11. Let E be a Hausdorff locally convex space and C be a closed convex locally compact cone which contains no line. Then C has a compact base.

Proof. Let V be a closed convex neighbourhood of 0, in E, such that $C \cap V$ is compact. In $C \cap V$, 0 is an extreme point (otherwise C would contain a line). Then by Bourbaki II-7-1 Prop. 2 (or Meyer XI-9-c), every neighbourhood of 0 in $C \cap V$ contains a set $(C \cap V) \cap F$ where F is an open half space of E such that $0 \in F$. Let F be such an open half space such that $C \cap V \cap F \subset C \cap \frac{1}{2} V$, and let H be the closed hyperplane which delimits F. First remark that $C \cap F \subset C \cap V$: indeed if there exists $y \in (C \cap F) - (C \cap V)$, for a suitable $\lambda \in]0, 1[$, λy would belong to $C \cap V$ (and still to $C \cap F$) hence to $C \cap V \cap F$, but not to $C \cap \frac{1}{2} V$. Consider $x \in C - \{0\}$. The half line $\mathbb{R}_+ x$ is not contained in F (because $C \cap F$ is contained in the compact set $C \cap V$). Hence $\mathbb{R}_+ x$ meets H. That proves that $C \cap H$ is a base of C. Finally $C \cap H \subset \overline{C \cap F} \subset C \cap V$ proves that $C \cap H$ is compact.

§ 3 - INF-COMPACTNESS PROPERTIES -

12 - Let F be a Hausdorff topological space and $h : F \to \overline{\mathbb{R}}$. The function h is said to be inf-compact if for every $\alpha \in \mathbb{R}$ the set $\{y \in F \mid h(y) \le \alpha\}$ is compact. For brevity we shall write $\{h \le \alpha\}$ instead of $\{y \mid h(y) \le \alpha\}$. If F is a vector space and $y' \in F'$, the function h is said to be inf-compact for the slope y' if for every $\alpha \in \mathbb{R}$ the set $\{h - \langle y', . \rangle \le \alpha\}$ is compact. The meanings of "h is inf-equicontinuous" (if F is a dual space), "h is inf-complete" etc... are obvious.

Theorem I-12. Let E be a Hausdorff locally convex space and f a convex l.s.c. proper function on E. Then the following properties are equivalent :

a) f is finite and continous at x_o

b) there exists $\alpha \in \mathbb{R}$ such that

$\alpha > \inf \{f^*(x') - \langle x', x_o\rangle \mid x' \in E'\}$ and

$\{f^* - \langle ., x_o\rangle \le \alpha\}$ is equicontinuous

c) f^* is inf-equicontinuous for the slope x_o.

Proof.

1) First suppose the theorem is true for $x_o = 0$. Let $x_1 \in E$. Put $\tilde{f}(x) = f(x + x_1)$. By hypothesis the result is true for $\tilde{f}$, $\tilde{f}^*$ and 0. But from $f(x) = \tilde{f}(x - x_1)$ follows

$$\begin{aligned} f^*(x') &= \sup \{\langle x', x\rangle - f(x) \mid x \in E\} \\ &= \sup \{\langle x', x\rangle - \tilde{f}(x - x_1)\} \\ &= \sup \{\langle x', x_1 + y\rangle - \tilde{f}(y) \mid y \in E \} \\ &= \langle x', x_1\rangle + \tilde{f}^*(x'). \end{aligned}$$

So $\{\tilde{f}^* \le \beta\} = \{f^* - \langle ., x_1\rangle \le \beta\}$ and the theorem is true for any x_1.

2) Now we suppose $x_0 = 0$ and we prove $a \Rightarrow c$. Let V be a neighbourhood of 0, and $\lambda \in \mathbb{R}$ such that $f(x) \leq \lambda$ on V. Then

$$f \leq \lambda + \delta(.|V), \text{ and}$$

$$f^* \geq (\lambda + \delta(.|V))^*.$$

But $(\lambda + \delta(.|V))^*(x') = \delta^*(x'|V) - \lambda$.

Then for $\alpha \in \mathbb{R}$

$$\{f^* \leq \alpha\} \subset \{(\lambda + \delta(.|V))^* \leq \alpha\}$$

$$= \{x' | \delta^*(x'|V) \leq \alpha + \lambda\}$$

$$= (\alpha + \lambda)V^\circ \text{ if } \alpha + \lambda \geq 0$$

As $\{f^* \leq \alpha\}$ is increasing in α, all the sets $\{f^* \leq \alpha\}$ are equicontinuous.

3) As $c \Rightarrow b$ is obvious, it remains to prove $b \Rightarrow a$ with $x_0 = 0$. We suppose that $\alpha \in \mathbb{R}$, $\alpha > \inf\{f^*(x') | x' \in E'\}$ and $\{f^* \leq \alpha\}$ is equicontinuous. As f^* is l.s.c. the set $\{f^* \leq \alpha\}$ is $\sigma(E',E)$ closed, hence $\sigma(E',E)$ compact, and f^* is bounded below. Then

$$\inf f^*(x') = - f(0) \text{ is finite.}$$

If epi f^* is translated by (x'_0, r_0) the polar function becomes

$$\delta^*((x, -1) | \text{epi } f^* + (x'_0, r_0))$$

$$= \delta^*((x, -1) | \text{epi } f^*) + < x'_0, x > - r_0$$

$$= f(x) + < x'_0, x > - r_0.$$

So to prove continuity of f at 0 we may translate epi f^*. We shall suppose

$$\alpha > f^*(0) = 0.$$

Then $C = \{f^* \leq \alpha\}$ is equicontinuous convex closed and contains 0. Let j be the gauge of C. It is easy to see (by properties of convex functions of one variable) that

$\alpha j(x') \leq f^*(x')$, for every $x' \in E - C$,

and that

$\alpha j(x') - [\alpha - \inf_{y'} f^*(y')] \leq f^*(x')$, for every $x' \in C$.

Thus $\alpha j - [\alpha - \inf_{y'} f^*(y')] = \alpha j - (\alpha + f(0)) \leq f^*$

and

$$f \leq [\alpha j - (\alpha + f(0))]^*$$

But $[\alpha j - (\alpha + f(0))]^* = \alpha + f(0) + (\alpha j)^*$

$$= \alpha + f(0) + \delta(.|\alpha C^\circ)$$

(by lemma 6). Hence f is bounded above on αC° which is a neighbourhood of 0, and thus f is continuous by a well known theorem (Bourbaki II-2-10 Prop. 21, p. 60).

13 - Lemma I-13. Let E be a Hausdorff locally convex space and f a convex l.s.c. function on E. Then the following properties are equivalent :

a) epi f is locally compact.

b) for every $x' \in E'$ f is inf-locally compact for the slope x'

c) there exists $x' \in E'$ and $\alpha \in \mathbb{R}$ such that

$\alpha > \inf\{f(x) - < x', x > | x \in E\}$ and

$\{ f - < x', . > \leq \alpha\}$ is locally compact.

Proof.

1) First it is easy to see that

$$\text{epi}(f-x') = \{(x,r) \in E \times \mathbb{R} | r \geq f(x) - < x', x >\}$$

is equal to $\varphi(\text{epi } f)$, where

$\varphi : E \times \mathbb{R} \to E \times \mathbb{R}$ is defined by

$\varphi(x, r) = (x, r - < x', x >)$.

As φ is an isomorphism we have

"epi f is locally compact" $\Leftrightarrow$ "epi$(f-x')$ is locally compact"

2) We prove now a $\Rightarrow$ b. We may suppose $x' = 0$. Then

$$\{f \leq \alpha\} = pr_E(\text{epi } f \cap E \times \{\alpha\}),$$

where pr_E is the projection onto E, and is an homeomorphism from $E \times \{\alpha\}$ to E.

3) As b $\Rightarrow$ c is obvious it remains to prove c $\Rightarrow$ a. We may again suppose $x' = 0$. We may suppose without loss of generality that $0 \in$ epi f and $\alpha > 0$. Let $r \geq \alpha$. By a suitable homothety $(y, k) \to \lambda(y, k)$ with $\lambda \in]0, 1]$ such that $\lambda r = \alpha$, epi $f \cap E \times \{r\}$ is transformed into a closed subset of epi $f \cap E \times \{\alpha\}$ (thanks to convexity). Hence $\{f \leq r\}$ is locally compact. That is true for any r, because $r \mapsto \{f \leq r\}$ is increasing. Finally any $(x,\lambda) \in$ epi f admits locally compact neighbourhoods : if $r > \lambda$, $\{f \leq r\} \times]-\infty, r]$ is one neighbourhood.

14 - Theorem I-14. Let E be a Hausdorff locally convex space. Consider on E the Mackey topology $\tau(E, E')$ and on E' the weak topology $\sigma(E', E)$. Let f be a convex l.s.c. proper function on E. Then the following properties are equivalent :

a) there exists $x_0 \in E$ such that f is finite and continuous at x_0 (by theorem 12 that is equivalent to f^* is inf-compact for the slope x_0).

b) epi f^* is locally compact and contains no line.

c) for every $x \in E$ and $\alpha \in \mathbb{R}$, $\{f^* - \langle \cdot, x \rangle \leq \alpha\}$ is locally compact and contains no line.

d) <u>there exists</u> $x \in E$ <u>and</u> $\alpha \in R$ <u>such that</u>

$$\alpha > \inf\{f^*(x') - \langle x', x\rangle \mid x' \in E'\} \text{ and } \{f^* - \langle \cdot, x\rangle \leq \alpha\}$$

<u>is locally compact, and</u> epi f^* <u>contains no line</u>.

<u>Proof</u>. 1) The equivalence of b, c, d follows from lemma 13. And $a \Rightarrow d$ results from theorem 12 (if epi f^* contains a line the set $\{f^* - \langle \cdot, x_0\rangle \leq \alpha\}$ contains a half-line. But that is not possible since it is compact).

2) Now we shall prove $b \Rightarrow a$. The asymptotic cone $C =$ As(epi f^*) contains $\{0\} \times \mathbb{R}_+$. Lemma 11 applies to C : there exist $(x_0, \lambda) \in E \times \mathbb{R}$ and $r \in \mathbb{R}$ such that

$$S = C \cap \{(x', t) \mid \langle (x', t), (x_0, \lambda)\rangle = r\}$$

is a compact base of C. As $\{0\} \times \mathbb{R}_+ \subset C$, $\lambda \neq 0$. We may suppose $\lambda = -1$, and as $(0, \frac{r}{\lambda}) \in S$, $r < 0$, and we may suppose $r = -1$. Hence

$$S = C \cap \{(x', t) \mid t = 1 + \langle x', x_0\rangle\}.$$

Let $\alpha \in \mathbb{R}$ and $A = \{f^* - \langle \cdot, x_0\rangle \leq \alpha\}$. Then A is locally compact and contains no line. Suppose $y' \in As(A)$. Then for every $k \geq 0$ and $x'_0 \in A$, $x'_0 + k\, y' \in A$, that is $f^*(x'_0 + k\, y') \leq \langle x'_0 + k\, y', x_0\rangle + \alpha$ or $(x'_0, \langle x'_0, x_0\rangle + \alpha) + k(y', \langle y', x_0\rangle) \in$ epi f^*, that is $(y', \langle y', x_0\rangle) \in$ As (epi f^*) $= C$.

If $y' \neq 0$, there exists $k_0 > 0$ such that $k_0(y', \langle y', x_0\rangle) \in S$, that is $k_0 \langle y', x_0\rangle = 1 + \langle k_0 y', x_0\rangle$.

This is a contradiction. Thus $As(A) = \{0\}$. By theorem 9 A is compact. As the topology of E' is $\sigma(E', E)$, theorem 12 implies that f is $\tau(E, E')$ continuous at x_0.

15 - <u>Corollary I.15</u>. <u>Let</u> F <u>be a Hausdorff locally convex space and</u> C <u>a closed convex non empty subset of</u> F. <u>We consider on</u> F <u>the weak</u>

topology $\sigma(F, F')$ *and on* F' *the Mackey topology* $\tau(F', F)$. *Then the following properties are equivalent :*

a) *there exists* $y'_0 \in F'$ *such that* $\delta^*(.|C)$ *is finite and continuous at* y'_0

b) C *is locally compact and contains no line.*

c) *there exists* $y'_0 \in F'$ *such that for every* $\beta \in \mathbb{R}$

$\{y \in C | < y'_0, y > \geq \beta\}$ *is compact.*

Moreover a) *and* c) *are satisfied by the same* y'_0.

Proof. This follows from theorems 12 and 14 applied to $f = \delta^*(.|C)$ and $f^* = \delta(.|C)$.

Theorem 12 gives $a \Leftrightarrow c$, and theorem 14 gives $a \Leftrightarrow b$.

Remarks 1) It is possible to prove first corollary 15 and to deduce from it theorem 14. Indeed "f is finite and continuous at x_0" $\Leftrightarrow$ "$\delta^*(.|\text{epi } f^*)$ is finite and continuous at $(x_0, -1)$" (we recall that for $r < 0$, $\delta^*((x, r)|\text{epi } f^*) = - rf(- \frac{x}{r})$).

2) Joly [8] gives a characterization of f such that epi f^* is locally compact (and possibly contains lines) 3)

16 - *Theorem I-16*. *Let* E *and* F *be two Hausdorff locally convex spaces,* A *a linear map from* E *to* F *with transpose* A^* *from* F' *to* E'. *Let* C *be a closed (for a topology compatible with duality) convex subset of* F'. *Suppose* $\delta^*(.|C)$ *is finite and* $\tau(F, F')$ *continuous at some* $\bar{y} \in A(E)$. *Then* $A^*(C)$ *is closed.*

Proof. Let $(x'_i)_{i \in I}$ be a generalized sequence in $A^*(C)$ which converges weakly to $\bar{x}'$. Let $\bar{x} \in E$ such that $A(\bar{x}) = \bar{y}$. Let $y'_i \in C$ such that $A^*(y'_i) = x'_i$.

Then

$$< y'_i, \bar{y} > = < y'_i, A(\bar{x}) > = < A^*(y'_i), \bar{x} > = < x'_i, \bar{x} >.$$

As (x'_i) converges there exist $i_o \in I$ and $\alpha \in \mathbb{R}$ such that $< x'_i, \bar{x} > \geq \alpha$ for every $i \geq i_o$. Thus $< y'_i, \bar{y} > \geq \alpha$ for every $i \geq i_o$. By corollary 15 (or theorem 12) $\{y'_i | i \geq i_o\}$ is contained in a weak compact set. Let $\mathcal{U}$ be an ultrafilter on I finer than the filter of sections. Then $\lim_{\mathcal{U}} y'_i$ exists and belongs to C. As A^* is weakly continuous

$$A^*(\lim_{\mathcal{U}} y'_i) = \lim_{\mathcal{U}} A^*(y'_i) = \lim x'_i = \bar{x}'.$$

That proves $\bar{x}' \in A^*(C)$.

__Remark.__ Theorem 16 is related to theorems 22 and 29.

§ 4 - INFIMUM - CONVOLUTION -

17 - __Definition. Let__ E __be a vector space,__ f_1 __and__ f_2 __two functions from__ E __to__ $\bar{\mathbb{R}}$. __Then the infimum-convolution of__ f_1 __and__ f_2 __is the function denoted__ by $f_1 \nabla f_2$, __from__ E __to__ $\bar{\mathbb{R}}$, __defined by__ $(f_1 \nabla f_2)(x) = \inf\{f_1(x-y) \dotplus f_2(y) | y \in E\}$.

__Remark.__ It is clear that

$$(f_1 \nabla f_2)(x) = (f_2 \nabla f_1)(x)$$
$$= \inf\{f_1(x_1) \dotplus f_2(x_2) | x_1 \in E,\ x_2 \in E,\ x_1 + x_2 = x\}$$

The formula $f \nabla \delta(.|\{a\}) = f(. -a)$ justifies the terminology because an analogous formula holds for convolution of measures.

18 - Let E be a vector space and A a subset of $E \times \mathbb{R}$ such that $A + \{0\} \times [0, \infty[= A$. Then, for every $x \in E$, $A \cap \{x\} \times \mathbb{R}$ is an interval of $\{x\} \times \mathbb{R}$ unlimited on the right. So if E is endowed with the discrete topology, and $\mathbb{R}$ with the usual topology, the

closure of A in $E \times \mathbb{R}$ is the smaller epigraph which contains A. The function thus defined is also given by $f(x) = \inf\{r \in \mathbb{R} \mid (x, r) \in A\}$.

Proposition I-18. *Let* E, f_1 *and* f_2 *be as in definition 17. Then* $\text{epi}(f_1 \nabla f_2)$ *is the smaller epigraph which contains* $\text{epi } f_1 + \text{epi } f_2$. *Consequently if* f_1 *and* f_2 *are convex,* $f_1 \nabla f_2$ *is convex.*

Proof. First remark that $(\text{epi } f_1 + \text{epi } f_2) + \{0\} \times [0,\infty[\; = \text{epi } f_1 + \text{epi } f_2$.
By definition $(f_1 \nabla f_2)(x) = \inf\{f_1(x_1) \dotplus f_2(x_2) \mid x_1 + x_2 = x\}$.
But from

$$f_1(x_1) \dotplus f_2(x_2) = \inf\{r_1 + r_2 \mid (x_1, r_1) \in \text{epi } f_1, (x_2, r_2) \in \text{epi } f_2\}$$

follows

$$\begin{aligned}(f_1 \nabla f_2)(x) &= \inf\{r_1 + r_2 \mid (x_i, r_i) \in \text{epi } f_i, x_1 + x_2 = x\} \\ &= \inf\{r \mid (x,r) \in \text{epi } f_1 + \text{epi } f_2\}.\end{aligned}$$

19 - Proposition I-19. *Let* E *be a Hausdorff locally convex space and* f_1 *and* f_2 *two functions on* E. *Then*

$$(f_1 \nabla f_2)^* = f_1^* \dotplus f_2^*$$

(*with the convention* $(-\infty) + (+\infty) = -\infty$. *Remark that* f_i^* *takes the value* $-\infty$ *iff* f_i *is the constant* $+\infty$!).

Proof.

$$\begin{aligned}(f_1 \nabla f_2)^* (x') &= \delta^*((x', -1) \mid \text{epi}(f_1 \nabla f_2)) \\ &= \delta^*((x', -1) \mid \text{epi } f_1 + \text{epi } f_2) \\ &= \delta^*((x', -1) \mid \text{epi } f_1) \dotplus \delta^*((x', -1) \mid \text{epi } f_2)\end{aligned}$$

(the symbol $\dotplus$ is due to the fact that the support function of the empty set is $-\infty$)

$$= f_1^*(x') \dotplus f_2^*(x').$$

20 - Definition 20. We shall say that the infimum-convolution $f_1 \nabla f_2$ is exact at x, if $(f_1 \nabla f_2)(x) \in \mathbb{R}$ implies that there exist x_1 and x_2 such that $x_1 + x_2 = x$ and $f_1(x) + f_2(x) = (f_1 \nabla f_2)(x)$.

It is obvious that $f_1 \nabla f_2$ is exact at every point iff epi f_1 + epi f_2 is an epigraph. Another expected property is lower semicontinuity of $f_1 \nabla f_2$. If epi f_1 + epi f_2 is closed, by proposition 18, it is the epigraph of $f_1 \nabla f_2$ and so $f_1 \nabla f_2$ is exact and l.s.c. That leads to the following theorem.

21 - Theorem I-21. Let E be a Hausdorff locally convex space and f_1 and f_2 two convex l.s.c. proper functions. Suppose there exists $x_1' \in E'$ such that f_1^* and f_2^* are finite at x_1', and that f_1^* is continuous at x_1'. Then $f_1 \nabla f_2$ is exact and l.s.c.

Proof. By theorem 14 epi f is weakly locally compact, hence by n°20 and theorem 10 it suffices to prove

$$\mathrm{As}(\mathrm{epi}\ f_1) \cap (-\mathrm{As}(\mathrm{epi}\ f_2)) = \{0\}.$$

Suppose $(x, r) \in \mathrm{As}(\mathrm{epi}\ f_1) \cap (-\mathrm{As}(\mathrm{epi}\ f_2))$ and $(x, r) \neq 0$. By corollary 15 applied to epi f_1 and $(x_1', -1)$, $\langle (x_1', -1), (x, r) \rangle < 0$. Hence $\langle x_1', x \rangle < r$. And, since $f_2^*(x_1') = \delta^*((x_1', -1) | \mathrm{epi}\ f_2)$ is finite $\langle (x_1', -1), -(x, r) \rangle \leq 0$, hence $\langle x_1', x \rangle \geq r$. That contradicts $(x,r) \neq 0$.

Remarks. 1) It is possible to prove this theorem more directly. For example for exactness, there exist minimizing sequences such that $x_n + y_n = x$ and $f_1(x_n) + f_2(y_n) \to (f_1 \nabla f_2)(x)$. By a compactness arguments one can prove that (x_n) has a cluster point. Similarly for lower semi-continuity one can use generalized sequences.

2) For another result (using Choquet [4], see Lescarret [9].

22 - The following theorem is a consequence of theorem 16.

Theorem I-22. Let E and F be two Hausdorff locally convex spaces, and A a linear map from E to F with transpose A^* from F' to E'. Let f be a convex l.s.c. proper function on F, finite and $\tau(F, F')$ continuous at some $\bar{y} \in A(E)$. Then $(A^* \times 1_{\mathbb{R}})(\text{epi } f^*)$ is closed and is the epigraph of $(f \circ A)^*$.

Moreover

$$(f \circ A)^* (x') = \inf \{f^*(y') \mid y' \in F', A^*(y') = x'\}$$

and, if the infimum is finite, it is a minimum.

Proof. First $\delta^*(.\mid \text{epi } f^*)$ is continuous at $(\bar{y}, -1) \in (A \times 1_{\mathbb{R}})(E \times \mathbb{R})$. So $(A^* \times 1_{\mathbb{R}})(\text{epi } f^*)$ is closed by theorem 16. It is obvious that $\{0\} \times \mathbb{R}_+$ is contained in $(A^* \times 1_{\mathbb{R}})$ (epi f^*). So this set is an epigraph. We shall prove now that its polar function is $f \circ A$. Indeed :

$$\begin{aligned} \delta^*((x, -1) \mid (A^* \times 1_{\mathbb{R}})(\text{epi } f^*)) & \\ &= \sup \{<(x, -1), (A^* \times 1_{\mathbb{R}})(y', r)> \mid r \geq f^*(y')\} \\ &= \sup \{<(x, A^*(y')> - f^*(y') \mid y' \in F'\} \\ &= f^{**}(A(x)) = f(A(x)). \end{aligned}$$

Finally if $(f \circ A)^* (x') \in \mathbb{R}$

$$\begin{aligned} (f \circ A)^*(x') &= \min\{r \in \mathbb{R} \mid (x', r) \in \text{epi } (f \circ A)^*\} \\ &= \min\{r \in \mathbb{R} \mid (x', r) \in (A^* \times 1_{\mathbb{R}})(\text{epi } f^*)\} \\ &= \min\{f^*(y') \mid y' \in F', A^*(y') = x'\}. \end{aligned}$$

And if $(f \circ A)^*(x') = +\infty$ these formulas remain valid with inf in place of min (and the corresponding sets are empty).

23 - Application. We give another proof of theorem 21 when f_1^* and f_2^* are both finite and continuous at x_1'. This proof uses theorem 22.

Let $A : E' \to E' \times E'$ defined by

$A(x') = (x', x')$. The transpose is

$A^* : E \times E \to E$ defined by

$A^*(x, y) = x + y$.

Let $(f_1^* \oplus f_2^*)(x', y') = f_1^*(x') + f_2^*(y')$

and $(f_1 \oplus f_2)(x, y) = f_1(x) + f_2(y)$.

Then $f_1 \oplus f_2$ and $f_1^* \oplus f_2^*$ are convex l.s.c. proper functions on $E \times E$ and $E' \times E'$, which are mutually polar.

Then $f_1^* \oplus f_2^*$ is finite and continuous at $(x_1', x_1') \in A(E')$.

By theorem 22

$[(f_1^* \oplus f_2^*) \circ A]^* (x) = \inf \{(f_1^* \oplus f_2^*)^*(y, z) | A^*(y, z) = x\}$

and, if the infimum is finite, it is a minimum.

But $(f_1^* \oplus f_2^*) \circ A = f_1^* + f_2^*$ and $(f_1^* \oplus f_2^*)^* = f_1 \oplus f_2$. So

$$(f_1^* + f_2^*)^* (x) = \inf \{f_1(y) + f_2(z) | y + z = x\}$$
$$= (f_1 \nabla f_2)(x)$$

and, if the infimum is finite, it is a minimum.

Thus proposition 19 and theorem 21 are proved (but under a stronger hypothesis).

24 - In this paragraph we give an example where the infimum-convolution is l.s.c. but is not exact. The following property is useful to study lexicographic maximums (see Valadier [13] and Wegmann [14])

Proposition I-24. Let E be a Hausdorff locally convex space, C a non empty closed convex set, which is $\sigma(E, E')$ locally compact and contains no line. Suppose (see corollary 15) that $\delta^*(.|C)$ is finite

<u>and continuous at</u> x'_0. <u>Let</u> $\alpha \in \mathbb{R}$ <u>and let</u> H <u>be the hyperplane</u>

$H = \{x \in E \mid \langle x'_0, x\rangle = \alpha\}$

<u>Then</u> $\delta^*(x' \mid C \cap H) = [\delta^*(.\mid C) \triangledown \delta^*(.\mid H)](x') = \inf\{\delta^*(x' - \lambda x'_0 \mid C) + \lambda\alpha \mid \lambda \in \mathbb{R}\}$

<u>Proof.</u> Remark that $\delta(.\mid C \cap H) = \delta(.\mid C) + \delta(.\mid H)$.

Thus $\delta^*(.\mid C \cap H) = [\delta(.\mid C) + \delta(.\mid H)]^*$. By proposition 19 if $\delta^*(.\mid C) \triangledown \delta^*(.\mid H)$ belongs to $\Gamma(E')$ it is equal to $\delta^*(.\mid C \cap H)$.

Remark that

$$\delta^*(x' \mid H) = \begin{cases} \lambda\alpha & \text{if } x' = \lambda x'_0 \\ +\infty & \text{otherwise.} \end{cases}$$

Let $x' \in E'$ and let us prove that $\delta^*(.\mid C) \triangledown \delta^*(.H)$ is bounded above on a neighbourhood of x' (so by Bourbaki [2] II-2 n°10, Prop. 21, p. 60 $\delta^*(.\mid C) \triangledown \delta^*(.H)$ will be continuous. That is true even if $\delta^*(.\mid C) \triangledown \delta^*(.\mid H)$ is $-\infty$ at x'). Let V be an open neighbourhood of 0 such that $\delta^*(.\mid C)$ is bounded above by $M \in \mathbb{R}$ on $x'_0 + V$. Let $\lambda > 0$ such that $x' \in \lambda V$. Then for every $y' \in \lambda V$ one has

$$\begin{aligned} [\delta^*(.\mid C) \triangledown \delta^*(.\mid H)]\,(y') &\leq \delta^*(\lambda x'_0 + y' \mid C) + \delta^*(-\lambda x'_0 \mid H) \\ &\leq \lambda M + \delta^*(-\lambda x'_0 \mid H) \\ &= \lambda(M - \alpha). \end{aligned}$$

Finally a convex continuous function belongs to $\Gamma(E')$.

<u>Remark.</u> As we have proved

$$\text{int}\,(\text{dom}\,\delta^*(.\mid C)) + \text{dom}\,\delta^*(.\mid H) = E',$$

this proposition is a consequence of Moreau [11] 9. b. p. 56.

<u>Example.</u> (which proves that, generally, $\delta^*(.\mid C) \triangledown \delta^*(.\mid H)$ is not exact).

Let $E = \mathbb{R}^2$, $C = \{(x, y) \mid x^2 + y^2 \leq 1\}$, $H = \{(x, y) \mid x = 1\}$

Then $C \cap H = \{(1,0)\}$,

$$\delta^*((x', y') \mid C) = \sqrt{x'^2 + y'^2}$$

$$\delta^*((x', y') \mid H) = \begin{cases} x' & \text{if } y' = 0 \\ +\infty & \text{if } y' \neq 0, \end{cases}$$

$$[\delta^*(.|C) \nabla \delta^*(.|H)](x', y') =$$
$$= \inf \{\delta^*((x_1', y')|C) + \delta^*((x_2', 0)|H) | x_1' + x_2' = x'\}$$
$$= \inf \{\sqrt{x_1'^2 + y'^2} + x' - x_1' | x_1' \in \mathbb{R}\}.$$

The infimum is x'. If $y' = 0$ it is obtained for every $x_1' \geq 0$. If $y' \neq 0$ it is only the limit for $x_1' \to +\infty$. So the infimum-convolution is not exact.

§ 5 - SUB-DIFFERENTIABILITY -

25 - Definition. Let E be a topological vector space, $f : E \to \bar{\mathbb{R}}$, and $x_0 \in E$ such that $f(x_0) \in \mathbb{R}$. Then $x' \in E'$ is said to be a subgradient of f at x_0 if for every $x \in E$, $f(x) - f(x_0) \geq \langle x', x-x_0 \rangle$.

The set of all subgradients of f at x_0 is called subdifferential and is denoted by $\partial f(x_0)$.

Remarks. 1) If $f : E \to]-\infty, \infty]$ is not the constant $+\infty$, one can define a subgradient x' at x_0 by the formula

$$\forall x, \ f(x) \geq f(x_0) + \langle x', x - x_0 \rangle.$$

Then $f(x_0) = +\infty \Rightarrow \partial f(x_0) = \emptyset$.

2) For a convex function $f : E \to \bar{\mathbb{R}}$ and x_0 such that $f(x_0) \in \mathbb{R}$, one can define the directional derivative of f at x_0 in direction h, by

$$f'(x_0, h) = \inf \{\frac{1}{\lambda} [f(x_0+\lambda h) - f(x_0)] | \lambda > 0\}.$$
$$= \lim_{\lambda>0, \lambda \to 0} \frac{1}{\lambda} [f(x_0 + \lambda h) - f(x_0)].$$

It is clear that $x' \in \partial f(x_0)$ if and only if

$$\forall h \in E, \ \langle x_0', h \rangle \leq f'(x_0, h).$$

26 - Proposition I-26. 1) Let E be a Hausdorff locally convex space, $f : E \to \bar{\mathbb{R}}$, $x_0 \in E$ such that $f(x_0) \in \mathbb{R}$ and $x' \in E'$. Then the following properties are equivalent :

a) $x' \in \partial f(x_0)$

b) $f^*(x') + f(x_0) = < x', x_0 >$

c) $f^*(x') + f(x_o) \leq < x', x_o >$.

Consequently $\partial f(x_0)$ is closed and convex.

2) Let F be a Hausdorff locally convex space, C a closed convex non empty subset of F, and $y'_0 \in F'$ such that $\delta^*(y'_0|C) \in \mathbb{R}$. Then $\partial(\delta^*(.|C))(y'_0) = \{y \in C | <y'_o, y> = \delta^*(y'_0|C)\}$.

Proof.

1) First $x' \in \partial f(x_0) \Leftrightarrow < x', x_0 > - f(x_0) \geq < x', . > - f(.)$

$$\Leftrightarrow < x', x_0 > - f(x_0) \geq f^*(x')$$

$$\Leftrightarrow < x', x_0 > \geq f^*(x') + f(x_0).$$

That proves $a \Leftrightarrow c$. As for every $x' \in E'$, $f^*(x') \geq <x', x_o> - f(x_0)$ we have $b \Leftrightarrow c$.

Finally

$$\partial f(x_0) = \{x' \in E' | - <x', x_o> + f^*(x') \leq -f(x_0)\}$$

is $\sigma(E', E)$ closed and convex.

2) The second part is obvious from the first with $E = F'$, $E' = F$,

$f = \delta^*(.|C)$, $f^* = \delta(.|C)$, $x_0 = y'_0$. Indeed

$y \in \partial(\delta^*(.|C))(y'_0) \Leftrightarrow \delta(y|C) + \delta^*(y'_0|C) = < y'_0, y >$

(by 1)b))

$$\Leftrightarrow y \in C \text{ and } < y'_0, y > = \delta^*(y'_0|C).$$

27 - Theorem I-27. Let E be a Hausdorff locally convex space, f a convex l.s.c. proper function on E, and $x_0 \in E$ such that $f(x_0) \in \mathbb{R}$. Then $\partial f(x_0)$ is the projection on E' of the set of points of epi f^* which maximize $(x_0, -1)$.

Consequently if f is finite and continuous at any point x_1, then $\partial f(x_0)$ is weakly locally compact and contains no line.

<u>If in addition</u> f <u>is continuous at</u> x_0 <u>then</u> $\partial f(x_0)$ <u>is non-empty weakly compact and moreover for every</u> $h \in E$ <u>the directional derivative</u> $f'(x_0, h)$ <u>is equal to</u> $\max\{< x', h > | x' \in \partial f(x_0)\}$.

<u>Proof</u>.

1) We have

$$x' \in \partial f(x_0) \Leftrightarrow f^*(x') + f(x_0) = < x', x_0 >$$
$$\Leftrightarrow f^*(x') + \delta^*((x_0, -1)|\text{epi } f^*) = < x', x_0 >$$
$$\Leftrightarrow \delta^*((x_0, -1)|\text{epi } f^*) = <(x', f^*(x')), (x_0, -1)>.$$

Remark that if (x', r) belongs to epi f^* and maximizes $(x_0, -1)$, then $r = f^*(x')$. So the first part is proved.

If f is finite and continuous at x_1, then by theorem 14 epi f^* is locally compact and contains no line. The **surface** of epi f^* formed by points which maximize $(x_0, -1)$ is contained in the hyperplane $\{(x', r)|<(x', r), (x_0, -1)> = f(x_0)\} = \{(x', r)|r = <x', x_0> - f(x_0)\}$. But the projection of this hyperplane on E' is a homeomorphism. So $\partial f(x_0)$ is locally compact and contains no line.

2) We suppose now that f is continuous at x_0. By corollary 15 the surface of epi f^* in direction $(x_0, -1)$ is weakly compact and non-empty. There are many ways to prove the formula

$$f'(x_0, h) = \max\{< x', h > | x' \in \partial f(x_0)\}.$$

By the analytical Hahn-Banach theorem there exists $x^* \in E^*$ such that $< x^*, h> = f'(x_0, h)$ (note that $f'(x_0, .)$ is sublinear). As $f'(x_0, .) \leq f(x_0 + .)$, it is continuous, and so x^* belongs to E'. As $< x', h > \leq f'(x_0, h)$ for every $x' \in \partial f(x_0)$ the formula is proved.

Remarks. 1) A sophisticated proof of the formula would be to use theorem 28 (which does not use the required formula). Consider $g = \delta(.|\{x_0+\lambda h | \lambda \in \mathbb{R}\})$. Let $z' \in E'$ be such that $< z', h > = f'(x_0, h)$. Then $z' \in \partial(f+g)(x_0)$ and by theorem 28 $z' = x' + y'$ with $x' \in \partial f(x_0)$ and $y' \in \partial g(x_0)$. As $< y', h > = 0$ the formula is proved.

2) If C is a closed convex set with non empty interior

$$\partial(\delta(.|C))(x_0) = \{x' | \forall\, x \in C,\ < x', x-x_0 > \leq 0\}$$

is a closed convex cone, locally compact and which contains no line. If E is a Hilbert space, it is the cone of external orthogonal directions to C at x_0.

28 - Theorem I-28. Let E be a Hausdorff locally convex space, f_1 and f_2 two convex l.s.c. proper functions, and $x_1 \in E$ such that f_1 and f_2 are finite at x_1 and f_1 is continuous at x_1. Then for every $x \in E$,

$$\partial(f_1 + f_2)(x) = \partial f_1(x) + \partial f_2(x).$$

Proof. In the proof of theorem 21 we have proved (using theorem 10) that epi $f_1^* +$ epi f_2^* is closed so it is equal to $\text{epi}(f_1^* \nabla f_2^*)$ (and then to $\text{epi}(f_1 + f_2)^*$; indeed by proposition 19 $(f_1 + f_2) = (f_1^* \nabla f_2^*)^*$, and as $f_1(x_1) + f_2(x_1) < \infty$, $f_1^* \nabla f_2^*$ cannot take the value $-\infty$. Therefore $f_1 + f_2$ and $f_1^* \nabla f_2^*$ are mutually polar).

Remark that if A and B are two subsets of a vector space F, and if y' is a linear form on F, then

$$\{y | y \in A + B,\ y \text{ maximizes } y'\} = \{y_1 \in A | y_1 \text{ maximizes } y'\} + \{y_2 \in B | y_2 \text{ maximizes } y'\}.$$

So

$$pr_{E'}\{(x', r) \in epi(f_1 + f_2)^* | (x', r) \text{ maximizes } (x, -1)\}$$

$$= pr_{E'}\{(x_1', r_1) \in epi\ f_1^* | (x_1', r_1) \text{ maximizes } (x, -1)\}$$

$$+ pr_{Ee}\{(x_2', r_2) \in epi\ f_2 | (x_2', r_2) \text{ maximizes } (x, -1)\}$$

By theorem 27 that proves the theorem.

Remark. For another result see Lescarret [9].

29 - Theorem I-29. *Let* E *and* F *be two Hausdorff locally convex spaces, and* A *a linear map from* E *to* F *with transpose* A^* *from* F' *to* E'. *Let* f *be a convex l.s.c. proper function on* F, *finite and* $\tau(F, F')$ *continuous at some* $\bar{y} \in A(E)$.

Then for every $x \in E$

$$\partial(f \circ A)(x) = A^*(\partial f(Ax)).$$

Proof. By theorem 27

$$\partial f(Ax) = \{y' \in F' | \exists\ r \in \mathbb{R} \text{ such that } (y', r) \in epi\ f^* \text{ and maximizes } (Ax, -1)\}$$

and

$$\partial(f \circ A)(x) = \{x' \in E' | \exists\ r \in \mathbb{R} \text{ such that } (x', r) \in epi\ (f \circ A)^* \text{ and maximizes } (x, -1)\}$$

So

$$A^*(\partial f(Ax)) = \{x' | \exists (y', r) \in epi\ f^*, \text{ which maximizes } (Ax, -1) \text{ and such that } x' = A^*(y')\}$$

But $\langle (y', r), (Ax, -1)\rangle = \langle (A^*(y'), r), (x, -1)\rangle$.

Hence

$$A^*(\partial f(Ax)) = \{x' | \exists (y', r) \in epi\ f^* \text{ such that } (A^*(y'), r) \text{ maximizes } (x, -1) \text{ and such that } x' = A^*(y')\}.$$

As $(A^* \times 1_{\mathbb{R}})(\text{epi } f^*) = \text{epi } (f \circ A)^*$ (by theorem 22), we have proved

$$A^*(\partial f(Ax)) = \partial(f \circ A)(x).$$

<u>Remark.</u> In spite of its geometric character the above proof is not very short. So we give also the usual proof. First it is easy to see that $A^*(\partial f(Ax)) \subset \partial(f \circ A)(x)$.

Conversely if $x' \in \partial(f \circ A)(x)$, by proposition 26 one has

$$(f \circ A)(x) + (f \circ A)^*(x') = \langle x', x \rangle.$$

So $(f \circ A)^*(x')$ is finite, and by theorem 22 there exists y' such that $(f \circ A)^*(x') = f^*(y')$ and $x' = A^*(y')$.

Thus $f(Ax) + f^*(y') = \langle A^*(y'), x \rangle$

$$= \langle y', A(x) \rangle.$$

Hence $x' = A^*(y')$ with $y' \in \partial f(Ax)$.

§6 - SOME EXAMPLES OF MUTUALLY POLAR FUNCTIONS.

30 - <u>General remark.</u> Let $f : \mathbb{R} \to]-\infty, \infty]$ be a l.s.c. convex proper function, and $x_0 \in \mathbb{R}$ such that $f(x_0)$ is finite. Then $\partial f(x_0) = [-f'(x_0,-1), f'(x_0,1)] \cap \mathbb{R}$ (possibly empty if $f'(x_0,1) = -f'(x_0,-1) = \pm\infty$). By theorem 27 f^* is affine on the interval $[-f'(x_0,-1), f'(x_0,1)] \cap \mathbb{R}$, with slope x_0 (that can be proved also by geometric arguments: recall that $f^*(x') = \delta^*((x',-1) | \text{epi } f))$.

So an angular point of f corresponds to an interval on which f^* is affine. For more general results see Asplund Rockafellar [1].

More generally one can say that sharpness of one of the functions f and f^* corresponds to flatness of the other one (recall that $f \leq g \Leftrightarrow f^* \geq g^*$) : that is illustrated by theorem 12, by the previous remark and by the following result : f is uniformly continuous on E if and only if dom f^* is equicontinuous (Moreau [11] 7.e. and 8.h.).

31 - <u>Examples</u> 1) Let $E = \mathbb{R}$, $f(x) = \begin{cases} \frac{1}{x} & \text{if } x > 0 \\ +\infty & \text{if } x \leq 0 \end{cases}$

Then $f^*(x') = \begin{cases} -2\sqrt{-x'} & \text{if } x' \leq 0 \\ +\infty & \text{if } x' > 0. \end{cases}$

If $g(x) = \begin{cases} -\sqrt{x} & \text{if } x \geq 0 \\ +\infty & \text{if } x < 0 \end{cases}$ then $g^*(x') = \begin{cases} -\frac{1}{4x'} & \text{if } x' < 0 \\ +\infty & \text{if } x' \geq 0 \end{cases}$

2) Let $E = \mathbb{R}$, $p \in]1,\infty[$ $f(x) = \frac{1}{p}|x|^p$. Then if q is the conjugate of $p(\frac{1}{p} + \frac{1}{q} = 1)$, $f^*(x') = \frac{1}{q}|x'|^q$.

If $g(x) = |x|$, then $g^*(x') = \delta(x'|[-1,1])$

$$= \begin{cases} 0 & \text{if } x' \in [-1,1] \\ +\infty & \text{if } x' \notin [-1,1] \end{cases}$$

32 - <u>Young functions</u>. These functions are used in definition of Orlicz spaces and to obtain mutually polar functions (see next paragraph).

<u>Definition</u>. <u>We shall say that a function</u> $f : \mathbb{R} \to \overline{\mathbb{R}}$ <u>is a Young function if</u> f <u>is l.s.c. convex, proper and moreover</u> $\forall x \in \mathbb{R}$, $f(-x) = f(x)$.

<u>Properties</u>

1) $\forall x \in \mathbb{R}$, $f(0) \leq f(x)$ (because $0 = \frac{1}{2}[x + (-x)]$) and $f(0) \in \mathbb{R}$.

2) $x \mapsto f(x)$ is increasing on $[0,\infty[$.

3) f^* is a Young function and if $x' \geq 0$, $f^*(x') = \sup\{x'x - f(x) | x \geq 0\}$

4) Put $f(\pm\infty) = \lim_{x \to \infty} f(x)$. Then

$f(\pm\infty)$ is finite $\Leftrightarrow$ f is a finite constant $c \Leftrightarrow f^* = \delta(.|\{0\}) - c$.

$(c \in \mathbb{R})$

5) $f(\pm\infty) = +\infty \Leftrightarrow f$ is inf-compact

$\Leftrightarrow f^*$ is continuous at 0.

6) $f_{|[0,\infty]}$ is continuous on the left of every point

7) If $g : [0,\infty[\to]-\infty,\infty]$ has the following properties : $g(0) \in \mathbb{R}$, g is increasing, convex and continuous on the left of every point, then the function $f = g(|.|)$ is a Young function.

33 - Let E be a Hausdorff locally convex space. Let $A \subset E$ and $B \subset E'$ be two mutually polar sets (that is $A^o = B$, $B^o = A$). Let φ and ψ be two mutually polar Young functions. We denote by j_A the gauge of A :

$j_A(x) = \inf\{\lambda > 0 | x \in \lambda A\}$ (see n°6).

$= \inf\{\lambda \geq 0 | x \in \lambda A\}$ (because $x \in 0\,A \Leftrightarrow x = 0$).

<u>Theorem</u> I-33. a) <u>If</u> $\varphi(\infty) = \psi(\infty) = +\infty$, <u>then</u> $\varphi \circ j_A$ <u>and</u> $\psi \circ j_B$ <u>are mutually polar</u>.

b) <u>If</u> φ <u>is a finite constant and if</u> $j_B(x') = 0$ <u>implies</u> $x' = 0$, <u>then</u> $\varphi \circ j_A$ <u>and</u> $\psi \circ j_B$ <u>are mutually polar</u>.

<u>Proof</u> a) First remark that

$\varphi(j_A(x)) = \inf\{\varphi(\lambda) | \lambda \geq 0,\ x \in \lambda A\}$ (infimum in the ordered set $\overline{\mathbb{R}}$)

Indeed : (i) if $j_A(x) = +\infty$, $\{\lambda \geq 0 | x \in \lambda A\}$ is empty and $\varphi(j_A(x)) = +\infty$.

(ii) if $j_A(x) \in]0,\infty[$, x belongs to $j_A(x)A$ and the formula results from the fact that φ is increasing on $[0,\infty]$.

(iii) if $j_A(x) = 0$, the formula holds because φ is increasing and continuous at 0 (remark that it may happen that $\{\lambda \geq 0 | x \in \lambda A\} =]0,\infty[$). Therefore

$(\varphi \circ j_A)^*(x') = \sup\{\langle x', x\rangle - \varphi(j_A(x)) | x \in E\}$

$= \sup\{<x',x> - \varphi(\lambda) | x \in E,\ \lambda \geq 0,\ x \in \lambda A\}$

$= \sup\{<x',\lambda a> - \varphi(\lambda) | a \in A,\ \lambda \geq 0\}$

$= \sup\{\lambda<x',a> - \varphi(\lambda) | a \in A_+,\ \lambda \geq 0\}$

(where $A_+ = \{a \in A | < x',a > \geq 0\}$)

$= \sup \{\psi(< x',a >) | a \in A_+\}$

(thanks to 3) of n°32)

$= \psi(j_B(x'))$ because

$\sup \{< x', a > | a \in A_+\} = \delta^*(x' | A) = j_B(x')$ by lemma 6, and $\psi|[0,\infty]$ is increasing and continuous on the left of every point.

b) If $\varphi = c \in \mathbb{R}$, and $j_B(x') = 0 \Rightarrow x' = 0$, one has

$\varphi \circ j_A = c$, and $\psi(j_B(x')) = \begin{cases} 0 & \text{if } x' = 0 \\ -c & \text{if } x' \neq 0. \end{cases}$

Hence $\varphi \circ j_A$ and $\psi \circ j_B$ are still mutually polar.

Applications

1) If $\varphi(x) = |x|$, $\varphi \circ j_A = j_A$ and

$$\psi \circ j_B(x') = \begin{cases} 0 & \text{if } j_B(x') \leq 1 \\ +\infty & \text{if } j_B(x') > 1 \end{cases}$$

$$= \delta(x' | B)$$

So theorem 33 contains as a special case the formula of lemma 6 (that is not surprising).

2) If E is a normed space, one can take for A and B the unit balls. Then $j_A(x) = \|x\|$, $j_B(x') = \|x'\|$. Therefore $x \mapsto \varphi(\|x\|)$ and $x' \mapsto \psi(\|x'\|)$ are mutually polar.

34 - In this paragraph we use theorem 22 to prove a theorem (whose direct proof is tedious) about quadratic form and trinomial on Hilbert space.

Let H be a real Hilbert space. Then by theorem 33 and n°31, 2) the function $f : x \mapsto \frac{1}{2}\|x\|^2$ is equal to its polar. Let A be a positive injective hermitian continuous operator on H. It is well known that there exists a positive hermitian operator $A^{\frac{1}{2}}$ such that $A^{\frac{1}{2}}A^{\frac{1}{2}} = A$, and $A^{\frac{1}{2}}$ is still injective. Moreover if $R(A)$ denotes the range of A, one has $R(A) \subset R(A^{\frac{1}{2}})$.

Theorem I.34. a) Let $q(x) = \frac{1}{2}(Ax|x)$. Then

$$q^*(y) = \begin{cases} \frac{1}{2}\|(A^{\frac{1}{2}})^{-1}(y)\|^2 & \text{if } y \in R(A^{\frac{1}{2}}) \\ +\infty & \text{if } y \notin R(A^{\frac{1}{2}}) \end{cases}$$

b) Let $b \in H$ and $t(x) = q(x) - (b|x)$. Then

(i) if $b \in R(A)$, t attains its infimum at $A^{-1}(b)$,

(ii) if $b \in R(A^{\frac{1}{2}}) - R(A)$, t is bounded below but does not attain its infimum,

(iii) if $b \notin R(A^{\frac{1}{2}})$, $\inf\{t(x) | x \in H\} = -\infty$.

Proof a) We apply theorem 22. We have $q = f \circ A^{\frac{1}{2}}$.
As f is continuous on H and $(A^{\frac{1}{2}})^* = A^{\frac{1}{2}}$, the formula of theorem 22 gives
$$q^*(y) = \inf\{\tfrac{1}{2}\|z\|^2 \,|\, A^{\frac{1}{2}}(z) = y\}.$$
That proves the required formula.

b) One has

$$t^*(y) = \sup\{(y|x) + (b|x) - q(x) | x \in E\}$$
$$= q^*(y+b).$$

Thus $\inf\{t(x)|x \in H\} = -t^*(0) = -q^*(b)$.

If $b \notin R(A^{\frac{1}{2}})$, $\inf\ t(x) = -\infty$.

If $b \in R(A^{\frac{1}{2}})$, $\inf\ t(x)$ is finite. As t is Fréchet differentiable, with $t'(x) = Ax - b$, the infimum is attained at x_0 if and only if $Ax_0 - b = 0$, that is if $b \in R(A)$ and $x_0 = A^{-1}(b)$.

Example. Let $H = \ell^2(\mathbb{N})$, (λ_n) a bounded sequence of strictly positive numbers, $A((x_n)) = (\lambda_n x_n)$. Then $A^{\frac{1}{2}}((x_n)) = (\sqrt{\lambda_n}\, x_n)$.

If there exists a subsequence (λ_{n_k}) which converges to 0, one has $R(A) \subsetneqq R(A^{\frac{1}{2}}) \subsetneqq H$.

BIBLIOGRAPHY OF CHAPTER I

1. ASPLUND, E. ROCKAFELLAR, R.T. gradients of convex functions. Trans. A.M.S. 139(1969) 443-467.

2. BOURBAKI, - Espaces vectoriels topologiques. Ch. I - II 2ième ed., Ch. III - IV - V 1ère ed.

3. CASTAING, Ch. - Quelques applications du théorème de Banach Dieudonné. Montpellier 1969 - 70, Publication N° 67.

4. CHOQUET, G. - Ensembles et cônes faiblement complets. C.R. Acad. Sci. Paris. 254 (1962) - 1908 - 1910.

5. DIEUDONNE, J. - Sur la séparation des ensembles convexes. Math. Annalen 163 (1966) - 1-3.

6. IOFFE, A.D. - TIHOMIROV (V.M.) - Duality of convex functions and extremum problems. Uspehi Mat. N. 23-6 (1968) - 51-116.

7. IOFFE, A.D. - LEVIN (V.L.) - Subdifferentials of convex functions - Trudi Moskov. Mat. Ob. 26 (1972) - 3-73.

8. JOLY, J.L. - Une famille de topologies et convergences sur l'ensemble des fonctionnelles convexes - Thèse Grenoble 1970.

9. LESCARRET, C. - Sur la sous-différentiabilité d'une somme de fonctionnelles convexes semi-continues inférieurement. C.R. Acad. Sc. Paris - 262 (1966) - 443-446.

10. MEYER, P.A. - Probabilités et potentiel - Hermann Paris 1966.

11. MOREAU, J.J. - Fonctionnelles convexes. Polycopié Collège de France 1966-67.

12. ROCKAFELLAR, R.T. - Convex Analysis - Princeton University Press (1970).

13. VALADIER, M. Contribution a l'Analyse Convexe - Thesis Paris (1970).

14. WEGMANN, R. - Der Wertebereich von Vektoringegralem. Z. Warschein... 14 (1970) - 203-238.

Chapter II

HAUSDORFF DISTANCE AND HAUSDORFF UNIFORMITY

The purpose of this chapter is still to give material related to the study of multifunctions.

We begin by pretending the Hausdorff metric (following Kuratowski). An interesting theorem, in view of measurable multi-functions (though its interest now is more historical) is theorem 10 (Dubbins) which was first published in Debreu [4]. It describes the Borel tribe of the Hausdorff topology.

We also give Hormander's theorem (th.18) which characterizes Hausdorff topology on convex bounded subsets of a topological vector space with help of their support functions. That permits us to give an embedding theorem (th.19) which allows us to consider some multi-functions as vector valued functions. Finally we give a few results about continuity. Upper semi-continuity is useful for the study of differential equations with multi-valued right-hand sides. Point wise convergence of support functions is used by Pallu de la Barrière [12] and Godet-Thobie in the definition of multi-measures.

§ 1 - THE SPACE OF CLOSED SUBSETS OF A METRIC SPACE -

1 - In all this paragraph X will be a metric space with the metric d. We do not assume $d(x,y) < \infty$.

Definition 1 - Let A and B be subsets of X, the excess of A over B is defined as

$$e(A, B) = \sup\{d(x, B) \mid x \in A\}$$ (the supremum is taken in $[0, \infty]$ so that $\sup \emptyset = 0$). The Hausdorff distance of A and B is

$$h(A, B) = \max(e(A, B), e(B, A)).$$

Elementary properties.

1) $e(A, \emptyset) = \infty$ if $A \neq \emptyset$.

$e(\emptyset, B) = 0$

2) $e(A, B) = 0 \Leftrightarrow A \subset \bar{B}$

$h(A, B) = 0 \Leftrightarrow \bar{A} = \bar{B}$

3) $e(A, C) \leq e(A, B) + e(B, C)$

$h(A, C) \leq h(A, B) + h(B, C)$.

So the set $\mathcal{F}_f(X)$ of all closed subsets of X with the Hausdorff distance becomes a metric space.

Remarks. In $\mathcal{F}_f(X)$, $\emptyset$ is an isolated point. If d is bounded, so is h on $\mathcal{F}_f(X) - \{\emptyset\}$.

2 - Theorem II-2. If $A_n \to A$ in the metric space $\mathcal{F}_f(X)$, then

$$A = \bigcap_n \overline{\bigcup_{m \geq n} A_m}$$

$$= \bigcap_{\varepsilon > 0} \bigcup_n \bigcap_{m \geq n} B(A_m, \varepsilon)$$

$$= \bigcap_{W \in \mathcal{W}} \bigcup_n \bigcap_{m \geq n} W(A_m)$$

(In these formulas

$$B(A_m, \varepsilon) = \{x \in X \mid d(x, A_m) \le \varepsilon\}$$

$\mathcal{W}$ is the set of all entourages of the uniform structure of X and

$$W(A_m) = \{y \in X \mid \exists\, x \in A_m \text{ such that } (x, y) \in W\}).$$

Proof. 1) Let $B = \bigcap_n \overline{\bigcup_{m \ge n} A_m}$. Let $\varepsilon > 0$, $n \in N$ and $x \in A$. There exists $m \ge n$ such that $h(A_m, A) \le \varepsilon$, hence $d(x, A_m) \le \varepsilon$, and there exists a point $x_m \in A_m$ such that $d(x, x_m) \le 2\varepsilon$. Therefore $x \in \overline{\bigcup_{m \ge n} A_m}$ for each n. That proves $A \subset B$.

Let $x \in B$ and let us prove that $A_n \to A \cup \{x\}$ (that will prove $B \subset A$). From $A_n \to A$, follows

$e(A_n, A \cup \{x\}) \to 0$. Moreover we shall prove that

$e(A \cup \{x\}, A_n) = \max(e(A, A_n), d(x, A_n)) \to 0$.

It is sufficient to prove $d(x, A_n) \to 0$. Let p be such that: $m, n \ge p$ implies $h(A_n, A_m) \le \varepsilon$.

From $x \in B$ it follows that there exists $m \ge p$ such that

$d(x, A_m) \le \varepsilon$. Hence if $n \ge p$

$d(x, A_n) \le d(x, A_m) + h(A_m, A_n) \le 2\,\varepsilon$.

2) Let $B = \bigcap_{\varepsilon > 0} \bigcup_n \bigcap_{m \ge n} B(A_m, \varepsilon)$.

If $x \in A$, as $d(x, A_p) \to 0$, it is obvious that x belongs to B. Conversely if $x \in B$, for each $\varepsilon > 0$, there exists n such that for each $m \ge n$,

$d(x, A_m) \le \varepsilon$, hence $e(A \cup \{x\}, A_n) \to 0$. And it is obvious that $e(A_n, A \cup \{x\}) \to 0$. Hence $h(A_n, A \cup \{x\}) \to 0$ and $A \cup \{x\} = A$.

3) The third equality is obvious because a basis of neighbourhoods is the family $W_\varepsilon = \{(x, y) \mid d(x, y) \le \varepsilon\} (\varepsilon > 0)$, and

$W_\varepsilon(A_n) \subset B(A_m, \varepsilon) \subset W_{2\varepsilon}(A_m)$.

3 - Theorem II-3. If X is a complete metric space, so is $\mathcal{S}_f(X)$.

Proof. Let (A_n) be a Cauchy sequence.

1) First remark that there exists N such that $n \geq N$, $m \geq N$ implies $h(A_n, A_m) \leq 1$. Then, either for every $n \geq N$, A_n is empty, or for every $n \geq N$, A_n is non empty. In the first case the sequence (A_n) converges to $\emptyset$. Let us consider the second case.

2) We shall prove that $\bigcap_n \overline{\bigcup_{m \geq n} A_m}$ is non empty.

Let $\varepsilon > 0$ (this will be useful later in 3). Now $\varepsilon = 1$ is sufficient). For each integer k there exists N_k such that $n \geq N_k$, $m \geq N_k$ implies $h(A_n, A_m) < 2^{-k}\,\varepsilon$. Let (n_k) be a strictly increasing sequence such that $n_k \geq N_k$. Let $x_0 \in A_{n_0}$. Suppose we have choosen $x_0, \ldots, x_k$ with the properties $x_i \in A_{n_i}$, $d(x_i, x_{i+1}) < 2^{-i}\,\varepsilon$. Then x_{k+1} is chosen in $A_{n_{k+1}}$ in order to satisfy $d(x_k, x_{k+1}) < 2^{-k}\,\varepsilon$ (this can be obtained for $d(x_k, A_{n_{k+1}}) \leq h(A_{n_k}, A_{n_{k+1}}) < 2^{-k}\,\varepsilon$).

The sequence (x_k) is Cauchy. Denote its limit by x. Then $x \in \bigcap_n \overline{\bigcup_{m \geq n} A_m}$.

3) The point x obtained in part 2) satisfies $d(x_0, x) \leq 2\,\varepsilon$. Therefore for every $n_0 \geq N_0$ and $x_0 \in A_{n_0}$ there exists a point $x \in A$ (here $A = \bigcap_n \overline{\bigcup_{m \geq n} A_m}$) such that $d(x_0, x) \leq 2\,\varepsilon$. Hence $e(A_{n_0}, A) \leq 2\,\varepsilon$ for $n_0 \geq N_0$.

4) Let us show now $e(A, A_n) \to 0$. This with 3), will prove that $h(A_n, A) \to 0$. Let $\varepsilon > 0$ and N be such that $m \geq N$, $n \geq N$ implies $h(A_n, A_m) \leq \varepsilon$. Let $x \in A$. Then $x \in \overline{\bigcup_{m \geq N} A_m}$. There exists $n_0 \geq N$ and $y \in A_{n_0}$ such that $d(x, y) \leq \varepsilon$. For each $m \geq N$ one has $d(x, A_m) \leq d(x, A_{n_0}) + h(A_{n_0}, A_m) \leq 2\,\varepsilon$. Hence $e(A, A_m) \leq 2\,\varepsilon$.

4 - Theorem II-4. Let $\mathcal{P}_{tb}(X)$ denote the set of all closed totally bounded subsets of X. Then $\mathcal{P}_{tb}(X)$ is closed in $\mathcal{P}_f(X)$.

Proof. Let (A_n) be a sequence in $\mathcal{P}_{tb}(X)$ which converges to $A \in \mathcal{P}_f(X)$. For $\varepsilon > 0$ there exists n such that $e(A, A_n) < \varepsilon$ and $x_1, \ldots, x_p$ such that the balls with centers x_i and radius ε cover A_n. Then the balls with centers x_i and radius 2ε cover A.

Remark. One can easily see that if X is totally bounded, then $\mathcal{P}_f(X)$ is totally bounded. Indeed if $\varepsilon > 0$ is given, let $x_1, \ldots, x_n$ be such that the open balls with centers x_i, and radius ε cover X. Let $A \in \mathcal{P}_f(X)$ and $I = \{i \mid B(x_i, \varepsilon) \cap A \neq \emptyset\}$. Then the set $B = \{x_i \mid i \in I\}$ has the property $h(A, B) \leq \varepsilon$. As the set of subsets of $\{x_1, \ldots, x_n\}$ is finite that proves that $\mathcal{P}_f(X)$ is totally bounded. Consequently if X is compact, $\mathcal{P}_f(X)$ is compact.

5 - Theorem II-5. If X is complete, the space $\mathcal{P}_k(X)$ of all compact subsets of X is complete.

Proof. This is obvious from theorems 3 and 4.

Remark. Theorem 5 remains valid if X is a uniform space (See Saint-Raymond [15]).

6 - Theorem II-6. The Hausdorff topology on the space of all compact subsets of X, $\mathcal{P}_k(X)$, is generated by the sets $\{K \in \mathcal{P}_k(X) \mid K \subset U\}$ (U open) and $\{K \in \mathcal{P}_k(X) \mid K \cap V \neq \emptyset\}$ (V open). A basis of neighbourhoods of K_0 consists of the sets $\{K \mid K \subset U, K \cap V_1 \neq \emptyset, \ldots, K \cap V_n \neq \emptyset\}$ (where $U, V_1, \ldots, V_n$ are open) which contain K_0.

Proof. 1) We shall prove that $\theta = \{K \mid K \subset U\}$ is open. Let $K_o \in \theta$. By compactness of K_o,

$$\varepsilon = \inf \{d(x, y) \mid x \in K_o, y \in X - U\} > 0.$$

Then $h(K, K_o) < \varepsilon \Rightarrow e(K, K_o) < \varepsilon \Rightarrow K \subset U$, that is $K \in \theta$.

We shall prove now that $\mathcal{U} = \{K \mid K \cap V \neq \emptyset\}$ is open. Let $K_o \in \mathcal{U}$. There exists an open ball with center $x_o \in K_o \cap V$ and radius ε which is contained in V. Then if $h(K, K_o) < \varepsilon$, K meets the ball, hence $K \cap V \neq \emptyset$ and $K \in \mathcal{U}$.

2) Conversely we shall prove that if $K_o \in \mathcal{P}_k(X)$ and $\varepsilon > 0$ are given, the ball of center K_o and radius ε contains a set $\{K \mid K \subset U\} \cap \{K \mid K \cap V_1 \neq \emptyset\} \cap \ldots \cap \{K \mid K \cap V_n \neq \emptyset\}$ which contains K_o. Indeed let $U = \{x \mid d(x, K_o) < \varepsilon\}$ and $V_1, \ldots, V_n$ be open balls of radius $\varepsilon/2$ which cover K_o. Then if $K \subset U$, $e(K, K_o) < \varepsilon$ and if K meets $V_1, \ldots, V_n$, $e(K_o, K) \le \varepsilon$.

Remark. Hence if T is a topological space, Γ, a multifunction from T to $\mathcal{P}_k(X)$, is continuous iff it is u.s.c. and ℓ.s.c. (see n°20, 21 for definitions of u.s.c. and ℓ.s.c.)

7 - Corollary II-7. If X is a metric space, the Hausdorff topology on the space of compact subsets of X, $\mathcal{P}_k(X)$, depends only on the topology of X (and not on the metric).

8 - Theorem II-8. If X is a separable metric space, so is $\mathcal{P}_k(X)$.

Proof. Let (x_n) be a dense sequence in X. Let $\mathcal{H}$ be the set of all finite sets $\{x_{i_1}, \ldots, x_{i_n}\}$. Then $\mathcal{H}$ is a countable part of $\mathcal{P}_k(X)$, and it is easy to verify (by theorem 6) that $\mathcal{H}$ is dense in $\mathcal{P}_k(X)$.

9 - Corollary II-9. If X is a Polish space, then $\mathcal{P}_k(X)$ with the topology described in theorem 6 is a Polish space.

10 - Theorem II-10. If X is a separable metric space, the Borel σ-field on $\mathcal{P}_k(X)$ (with the Hausdorff topology) is generated by the sets $\{K|K \subset U\}$ (U open) and is also generated by the sets $\{K|K \cap V \neq \emptyset\}$ (V open).

Proof. 1) Consider the set $\{K|K \cap V \neq \emptyset\}$. Remark that $V = \bigcup_n F_n$ with $F_n = \{x|d(x, X-V) \geq \frac{1}{n}\}$.

Then
$$\{K|K \cap V \neq \emptyset\} = \bigcup_n \{K|K \cap F_n \neq \emptyset\}$$
$$= \bigcup_n [\mathcal{P}_k(X) - \{K|K \subset X - F_n\}]$$
Hence the σ-field generated by the sets $\{K|K \cap V \neq \emptyset\}$ is included in the σ-field generated by the sets $\{K|K \subset U\}$.

2) Consider the set $\{K|K \subset U\}$. Put

$V_n = \{x|d(x, X - U) < \frac{1}{n}\}$. Then $X - U = \bigcap V_n$

Let us verify $K \cap (X - U) \neq \emptyset \Leftrightarrow \forall n,\ K \cap V_n \neq \emptyset$.

The implication $\Rightarrow$ is obvious. Conversely if $K \cap V_n \neq \emptyset$ for every n, let $x_n \in K \cap V_n$. Then any cluster point of the sequence (x_n) belongs to K and $X - U$. Hence

$$\mathcal{P}_k(X) - \{K|K \subset U\} = \bigcap_n \{K|K \cap V_n \neq \emptyset\}.$$

That proves that the σ-field generated by the sets $\{K|K \subset U\}$ is included in the σ-field generated by the sets $\{K|K \cap V \neq \emptyset\}$.

3) We shall now prove that any open set θ of $\mathcal{P}_k(X)$ belongs to the σ-field generated by the $\{K|K \subset U\}$ and the $\{K|K \cap V \neq \emptyset\}$. Indeed θ is the union of a family $\mathcal{A}$ of finite intersections of sets $\{K|K \cap U\}$ and $\{K|K \cap V \neq \emptyset\}$. (theorem 6). But as $\mathcal{P}_k(X)$ is

separable (theorem 8) θ is also the union of a countable subfamily of $\mathcal{A}$.

§ 2 - THE CASE OF A UNIFORM SPACE. HAUSDORFF UNIFORMITY.

11 - In this paragraph X is a Hausdorff uniform space, whose uniform structure is defined by a filtering family of semi-distances (or ecart) $(d_i)_{i \in I}$ (that is d_i is such that

$$d_i(x, y) \in [0, \infty]$$

$$d_i(x, y) = d_i(y, x)$$

$$d_i(x, z) \leq d_i(x, y) + d_i(y, z)).$$

Then the functions e_i and h_i defined as

$$e_i(A, B) = \sup\{d_i(x, B) \mid x \in A\}$$

and $h_i(A, B) = \max(e_i(A, B), e_i(B, A))$ have the properties :

$$e_i(A, \phi) = \infty \quad \text{if} \quad A \neq \phi$$

$$e_i(\phi, B) = 0$$

$$e_i(A, C) \leq e_i(A, B) + e_i(B, C)$$

$$h_i(A, C) \leq h_i(A, B) + h_i(B, C)$$

$$\forall i, \ e_i(A, B) = 0 \Leftrightarrow A \subset \bar{B}$$

$$\forall i, \ h_i(A, B) = 0 \Leftrightarrow \bar{A} = \bar{B}.$$

The family (h_i) is filtering.

Proof of the three last properties

1) First $\Leftarrow$ are obvious. Conversely if $\forall i$, $e_i(A, B) = 0$ and if $a \in A$, we have $\forall i$, $d_i(a, B) = 0$. Then any d_i-ball with strictly positive radius and center a meets B. Hence every neighbourhood of a meets B. Therefore $a \in \bar{B}$.

2) The last property follows from the fact that the correspondence $d \mapsto h$ is increasing.

12 - We examine now another way to define the uniform structure on $\mathcal{P}_f(X)$. Let $\mathcal{W}$ be a basis of entourages of the uniform structure of X. If $W \in \mathcal{W}$ we define $\tilde{W}$ by

$\tilde{W} = \{(A, B) \in \mathcal{P}_f(X)^2 | A \subset W(B)$ and $B \subset W(A)\}$ (we recall that $W(B) = \{y \in X | \exists\ x \in B$ such that $(x, y) \in W\})$.

<u>Theorem II-12. With the above notations, the set $\tilde{\mathcal{W}}$ of all $\tilde{W}$ is a basis of entourages on $\mathcal{P}_f(X)$. It defines the same uniform structure as the family of semi-distances</u> (h_i) (<u>see n°11</u>).

<u>Proof</u>. First the map $W \mapsto \tilde{W}$ is increasing. Then if $\mathcal{W}_0$ is another basis of entourages of X, each $\tilde{W}_0 \in \tilde{\mathcal{W}}_0$ contains a $\tilde{W} \in \tilde{\mathcal{W}}$ and vice versa. Now we consider $\mathcal{W}_0$ the set of all

$$U_{i,\varepsilon} = \{(x, y) \in X^2 | d_i(x, y) < \varepsilon\} \ (\varepsilon > 0,\ i \in I).$$

Then $e_i(A, B) < \varepsilon \Rightarrow A \subset U_{i,\varepsilon}(B) \Rightarrow e_i(A, B) < 2\varepsilon$

and $h_i(A, B) < \varepsilon \Rightarrow A \subset U_{i,\varepsilon}(B)$ and $B \subset U_{i,\varepsilon}(A) \Rightarrow h_i(A, B) < 2\varepsilon$

That proves that $\tilde{\mathcal{W}}_0$ is a basis of the uniform structure defined by the family (h_i).

<u>Remarks</u> 1) If X is an abelian topological group, let $\mathcal{V}$ be a basis of neighbourhoods of 0. Then the sets $\{(A, B) | A \subset B + V$ and $B \subset A + V\}$ $(V \in \mathcal{V})$ form a basis of entourages of the uniform structure of $\mathcal{P}_f(X)$.

If X is a Fréchet space $\mathcal{P}_f(X)$ is metrizable, but it can be more convenient to define the Hausdorff uniformity by the set of entourages

$\{(A, B) | A \subset B + V$ and $B \subset A + V\}$ with V a convex closed neighbourhood of 0 (or convex open neighbourhood).

2) Theorem 6 remains valid :

<u>Let</u> X <u>be a Hausdorff uniform space</u>. <u>The Hausdorff topology on</u> $\mathcal{P}_k(X)$ <u>is generated by the sets</u> $\{K \in \mathcal{P}_k(X) | K \subset U\}$ (U <u>open</u>) <u>and</u> $\{K | K \cap V \neq \emptyset\}$ (V <u>open</u>).

The proof remains valid, using the families (d_i), (e_i), (h_i), except for the first argument. It must be replaced by the following We shall prove that $\theta = \{K | K \subset U\}$ is open. Let $K_o \in \theta$. There exists an entourage W such that $W(K_o) \subset U$ (Bourbaki Top. Gén. ch. II. §4. prop.4 p.231). Thus there exists $i \in I$ and $\varepsilon > 0$ such that $\{x | d_i(x, K_o) < \varepsilon\} \subset U$. Then $h_i(K, K_o) < \varepsilon \Rightarrow e_i(K, K_o) < \varepsilon \Rightarrow K \subset U$, that is K belongs to θ.

§ 3 - THE SPACE OF CLOSED CONVEX SUBSETS OF A LOCALLY CONVEX SPACE -

13 - Let E be a Hausdorff locally convex vector space. Let $(p_i)_{i \in I}$ be a filtering family of semi-norms which defines the topology of E. Then $d_i(x, y) = p_i(x - y)$ is a semi-distance, and § 2 applies to E with the family $(d_i)_{i \in I}$.

<u>Theorem II-13</u>. <u>Let</u> $(F_\alpha)_{\alpha \in A}$ <u>be a generalized sequence of closed subsets of</u> E. <u>Suppose</u> (F_α) <u>converges to</u> F <u>with respect to the topology of</u> § 2. <u>Then if the</u> F_α <u>are convex so is</u> F, <u>if the</u> F_α <u>are bounded so is</u> F.

<u>Proof</u>. 1) Suppose the F_α are convex. Let x and $y \in F$, $\lambda \in [0, 1]$ and $z = \lambda x + (1-\lambda)y$. For every convex neighbourhood of 0, V, there exists α such that, for $\beta \geq \alpha$,

$$F \subset F_\beta + V \quad \text{and} \quad F_\beta \subset F + V.$$

Hence $F \cup \{z\} \subset F_\beta + V$ and $F_\beta \subset (F \cup \{z\}) + V$.

Therefore $F \cup \{z\}$ is also the limit of (F_α). That proves $z \in F$.

2) Suppose the F_α are bounded. For every convex neighbourhood of 0, V, there exists α such that $F \subset F_\alpha + V$. As F_α is bounded there exists $\lambda > 0$ such that $F_\alpha \subset \lambda V$, hence $F \subset (\lambda + 1)V$, and F is bounded.

Remark. If E is metrizable the first part follows from the last formula of theorem 2 : if $W = \{(x, y) | p_i(x - y) \leq \varepsilon\}$

$W(A_m)$ is convex, hence

$\bigcap_{m \geq n} W(A_m)$ is convex, and $\bigcup_n \bigcap_{m \geq n} W(A_m)$ is convex as it is the union of an increasing sequence of convex sets.

14 - Theorem II-14. If E is a Fréchet vector space, the following spaces with the metrizable Hausdorff uniformity are complete :

- the set of all closed convex sets
- the set of all closed bounded sets
- the set of all closed convex bounded sets
- the set of all compact convex sets.

Proof. This follows from theorems 3, 5 and 13.

15 - We recall the following (see I-2).

Definition. Let E be a Hausdorff locally convex vector space and A a subset of E. The support function of A is the function defined on E' by $x' \mapsto \delta^*(x'|A) = \sup\{\langle x', x\rangle | x \in A\}$.

16 - The following result is well known :

Theoem II-16. There is a one-to-one correspondence between non empty closed convex sets and sublinear $\sigma(E', E)$ lower semi-continuous functions on E' (with values in $]-\infty, \infty]$) which maps A into $\delta^*(.|A)$.

Proof. The support function $\delta^*(.|A)$ is sublinear, $\sigma(E', E)$ l.s.c., and $> -\infty$, when A is non empty. When furthermore A is closed and convex, $\delta^*(.|A)$ characterizes A, by the Hahn-Banach theorem. Finally every sublinear $\sigma(E', E)$ l.s.c. function φ is the support function of $A = \{x | \forall x' \in E', <x', x> \leq \varphi(x')\}$. This is a consequence of theorem I-4 : as φ is sublinear

$$\begin{aligned} 2\varphi^*(x) &= \sup\{<2x', x> - 2\varphi(x')\} \\ &= \sup\{<2x', x> - \varphi(2x')\} \\ &= \varphi^*(x) \end{aligned}$$

Hence for all x, $\varphi^*(x) = 0$ or ∞. The set

$$A = \{x | \varphi^*(x) = 0\} \text{ has support function } \varphi.$$

17 - Theorem II-17. Let A and B be closed non empty convex sets, and write $A \dotplus B = \overline{A + B}$. Then $\delta^*(.|A \dotplus B) = \delta^*(.|A) + \delta^*(.|B)$, and if $\lambda \in [0, \infty[$, $\delta^*(.|\lambda A) = \lambda\, \delta^*(.|A)$.

If A, B and C are closed convex bounded non empty sets then $A \dotplus C = B \dotplus C$ implies $A = B$.

Proof. The first part is obvious. Let us prove the second part. The support functions of A, B, C are finite valued. Then

$$\begin{aligned} \delta^*(x'|A) &= \delta^*(x'|A \dotplus C) - \delta^*(x'|C) \\ &= \delta^*(x'|B \dotplus C) - \delta^*(x'|C) \\ &= \delta^*(x'|B), \end{aligned}$$

and $A = B$.

18 - <u>Theorem II-18. Let</u> $\mathcal{P}_{cb}(E)$ <u>be the space of all convex closed bounded non empty subsets of</u> E. <u>Let</u> p <u>be a continuous semi-norm on</u> E, <u>and</u> U <u>the closed semi-ball</u> $U = \{x \mid p(x) \le 1\}$. <u>Let</u> e <u>be the excess and</u> h <u>be the Hausdorff semi-distance associated with</u> p. <u>Then</u>

$e(A,B) = \sup\{\delta^*(x'|A) - \delta^*(x'|B) \mid x' \in U^\circ\}$ <u>and</u>

$$h(A, B) = \sup\{|\delta^*(x'|A) - \delta^*(x'|B)| \mid x' \in U^\circ\}.$$

<u>Consequently the uniformity on</u> $\mathcal{P}_{cb}(E)$ <u>is defined by the family of semi-distances</u>

$$\sup\{|\delta^*(x'|A) - \delta^*(x'|B)| \mid x' \in K\} \quad (K \text{ equicontinuous set}).$$

<u>Proof</u>. Let $\bar{e}(A,B) = \sup\{\delta^*(x'|A) - \delta^*(x'|B) \mid x' \in U^\circ\}$.

Then, for $\varepsilon > 0$, $\bar{e}(A,B) \le \varepsilon$ is equivalent to

$$\forall x' \in E', \ \delta^*(x'|A) - \delta^*(x'|B) \le \varepsilon\, \delta^*(x'|U).$$

Indeed sufficiency is obvious. For necessity note that if $\delta^*(x'|U) < \infty$, then x' belongs to $\delta^*(x'|U)U^\circ$. But

$\forall x' \in E', \ \delta^*(x'|A) \le \delta^*(x'|B) + \varepsilon\delta^*(x'|U)$

is equivalent to

$$A \subset B \dot{+} \varepsilon U.$$

Finally $\inf\{\varepsilon > 0 \mid A \subset B \dot{+} \varepsilon U\} = e(A, B)$.

Indeed if $A \subset \overline{B + \varepsilon U}$ then $e(A, B) \le \varepsilon$ which entails the inequality $\ge$. And if $\varepsilon > e(A, B)$ then $A \subset \overline{B + \varepsilon U}$, so the inequality $\le$ holds.

<u>Remark</u>. That can also be proved using inf-sup theorem :

$$e(A,B) = \sup_{x \in A} \inf_{y \in B} p(x-y)$$

$$= \sup_{x \in A} \inf_{y \in B} \sup_{x' \in U^\circ} \langle x', x-y \rangle$$

$$= \sup_{x \in A} \sup_{x' \in U^o} \inf_{y \in B} \langle x', x-y \rangle$$

$$= \sup_{x \in A} \sup_{x' \in U^o} [\langle x', x \rangle - \delta^*(x'|B)]$$

$$= \sup_{x' \in U^o} [\delta^*(x'|A) - \delta^*(x'|B)].$$

19 - We consider now the problem of embedding $\mathcal{P}_{cb}(E)$ in a vector space.

Definition. Let $\mathcal{H}$ be the space of all real valued positively homogeneous functions, whose restrictions to equicontinuous sets K of E', are bounded and strongly continuous. With the topology of uniform convergence on equicontinuous sets, $\mathcal{H}$ becomes a Hausdorff locally convex vector space.

Theorem II-19. The space $\mathcal{H}$ is complete. The mapping from $\mathcal{P}_{cb}(E)$ to $\mathcal{H}$ defined by $i : A \to \delta^*(.|A)$ has the properties.

- it is injective
- $i(A \dotplus B) = i(A) + i(B)$
- $i(\lambda A) = \lambda i(A)$ $(\lambda \in [0, \infty[)$
- it is a homeomorphism from $\mathcal{P}_{cb}(E)$ onto its range.

Proof. First it is clear $\mathcal{H}$ is complete. We have to prove that $A \in \mathcal{P}_{cb}(E)$ implies $\delta^*(.|A) \in \mathcal{H}$. The non trivial fact is that $\delta^*(.|A)$ is bounded on each equicontinuous set. But that results from the fact that an equicontinuous set is strongly bounded (Bourbaki E.V.T. ch.III. prop 7 p.26). Finally, by theorem 16 i is injective, the two formulas result from theorem 17, and the last assertion from theorem 18.

Remark. If E is a Frechet space, so is $\mathcal{H}$. If E is a Banach space, then $\mathcal{H}$ with the norm

$$\|\varphi\| = \sup\{|\varphi(x')| \,|\, \|x'\| \leq 1\}$$

is Banach.

§ 4 - CONTINUITY OF CONVEX MULTIFUNCTIONS.

20 - Theorem II-20. Let T be a topological space, E a Hausdorff locally convex space, and Γ a multifunction from T to non empty subsets of E. We suppose $\Gamma(t_0)$ weakly compact and convex. Then Γ is weakly upper semicontinuous at t_0 if and only if the scalar functions $\delta(x'|\Gamma(.))$ are u.s.c. at t_0.

Remark. We say that Γ is u.s.c. at t_0 if for any open set U which contains $\Gamma(t_0)$, there exists a neighbourhood V of t_0 such that $t \in V$ implies $\Gamma(t) \subset U$.

Proof. 1) If Γ is u.s.c. at t_0 and $\alpha > \delta^*(x'|\Gamma(t_0))(\alpha \in \mathbb{R})$, put $U = \{x \in E | \langle x', x\rangle < \alpha\}$. There exists a neighbourhood V of t_0 such that $\Gamma(t) \subset U$ for every $t \in V$. Hence $\delta^*(x'|\Gamma(t)) \leq \alpha$ (this part remains true under weaker hypotheses than the given ones).

2) Suppose all the $\delta^*(x'|\Gamma(.))$ are u.s.c. If $\Gamma(t_0) = \emptyset$ then $\delta^*(0|\Gamma(t_0)) = -\infty$, and for t such that $\delta^*(o\,|\Gamma(t)) < 0$, $\Gamma(t) = \emptyset$. Hence Γ is u.s.c. at t_0. We suppose now $\Gamma(t_0) \neq \emptyset$. Let $x_0 \in \Gamma(t_0)$. Consider $\Gamma'(t) = \Gamma(t) - x_0$. We have
$\delta^*(x'|\Gamma'(t)) = -\langle x', x_o\rangle + \delta^*(x'|\Gamma(t))$.
That proves we may suppose $0 \in \Gamma(t_0)$. Let U be a weakly open set containing $\Gamma(t_0)$. There exists a closed convex neighbourhood of 0, V, such that $\Gamma(t_0) + V \subset U$ (Bourbaki Top. Gén. ch II §4 prop.4 p.231).

We can suppose V is the polar of a finite subset of E'. Because $\Gamma(t_0)$ is compact there exists $x_1,\dots,x_n \in \Gamma(t_0)$ such that the $x_i + \frac{1}{2}V$ cover $\Gamma(t_0)$. Let $A = co\{x_1,\dots,x_n\} + V$ (where co denotes the convex hull). Then A is closed and $A \subset U$. We may suppose $0 \in co\{x_1,\dots,x_n\}$ (because $0 \in \Gamma(t_0)$), then A^o is a finite dimensional convex polyhedral set contained in V^o :

$$A^o = co\{x'_1,\dots, x'_k\}.$$

From $\Gamma(t_0) \subset co\{x_1,\dots, x_n\} + \frac{1}{2}V \subset A \subset U$

follows $\delta^*(x'_j|\Gamma(t_0)) \leq \sup_i \delta^*(x'_j|x_i + \frac{1}{2}V) < \delta^*(x'_j|A) \leq 1.$

Let V be a neighbourhood of t_0 such that

$$\delta^*(x'_j|\Gamma(t)) \leq 1 \quad \text{for} \quad t \in V,\ j = 1,\dots, k.$$

Then $\Gamma(t) \subset A \subset U$ for every $t \in V$.

21 - Theorem II-21. *Let T be a topological space, E a Hausdorff locally convex space and Γ a multifunction from T to convex totally bounded subsets of E. We suppose $\bigcup_{t\in T} \Gamma(t)$ totally bounded. Then Γ is lower semicontinuous at t_0 if and only if the scalar functions $\delta^*(x'|\Gamma(.))$ are l.s.c. at t_0.*

Remark. We say that Γ is l.s.c. at t_0 if for any open set U which meets $\Gamma(t_0)$ there exists a neighbourhood V of t_0 such that $\Gamma(t) \cap U \neq \emptyset$ for every $t \in V$.

Proof. 1) Suppose Γ is l.s.c. at t_0. If $\alpha < \delta^*(x'|\Gamma(t_0))(\alpha \in \mathbb{R})$ then $\Gamma(t_0)$ meets the open set $U = \{x|\langle x', x\rangle > \alpha\}$. Then if t is in a neighbourhood of t_0, $\Gamma(t)$ meets U, and $\delta^*(x'|\Gamma(t)) > \alpha$. If $\delta^*(x'|\Gamma(t_0)) = -\infty$ (that is if $\Gamma(t_0) = \emptyset$) then $\delta^*(x'|\Gamma(.))$

is still l.s.c. at t_0 (this part remains true under weaker hypotheses that the given ones).

2) We suppose now all the $\delta^*(x'|\Gamma(.))$ are l.s.c. We may suppose $\Gamma(t_0) \neq \emptyset$(if $\Gamma(t_0) = \emptyset$, Γ is obviously l.s.c.). Let U be an open set which meets $\Gamma(t_0)$. As in theorem 20 we may suppose $0 \in \Gamma(t_0) \cap U$. We shall also suppose that U is a convex open set. If the theorem is false there exists a generalized sequence (t_α) which converges to t_0, such that $\Gamma(t_\alpha) \cap U = \emptyset$. By Hahn Banach there exists $x'_\alpha \in E'$ such that x'_α takes values ≤ -1 on $\Gamma(t_\alpha)$ and takes values ≥ -1 on U. Hence $x'_\alpha \in U^o$ (if exceptionally we define U^o as $\{x' | \forall\, x \in U, \langle x', x\rangle \geq -1\}$) and $\delta^*(x'_\alpha|\Gamma(t_\alpha)) \leq -1$. As U^o is equicontinuous it is compact for the topology of uniform convergence on totally bounded sets of E. Let z' be a cluster point of (x'_α) for this topology. From $0 \in \Gamma(t_0)$ follows $\delta^*(z'|\Gamma(t_0)) \geq 0$. For x'_β sufficiently near to z', from the fact that $\cup\,\Gamma(t)$ is totally bounded we have

$$\forall t, \qquad \delta^*(z'|\Gamma(t)) \leq \delta^*(x'_\beta|\Gamma(t)) + \frac{1}{2}.$$

But for each α there exists $\beta \geq \alpha$ such that x'_β belongs to a given neighbourhood of z'.

Hence $\delta^*(z'|\Gamma(t_\beta)) \leq \delta^*(x'_\beta|\Gamma(t_\beta)) + \frac{1}{2} \leq -\frac{1}{2}$.

This is not possible because $\delta^*(z'|\Gamma(.))$ is l.s.c. at t_0.

22 - Corollary II-22. Let T be a topological space, E a Hausdorff locally convex space, and Γ map T to convex compact non empty subsets of E. We suppose that each $t_0 \in T$ has a neighbourhood V such that $\bigcup_{t \in V} \Gamma(t)$ is contained in a compact set. Then if the support functions $\delta^*(x'|\Gamma(.))$ are continuous, Γ is continuous for the

Hausdorff topology.

Proof. Denote by K a compact set which contains $\bigcup_{t \in V} \Gamma(t)$. Let U be an open set. Then $U \cap K$ is also a weakly open set, and by theorem 20,

$\{t \in V | \Gamma(t) \subset U\}$ is open. If Θ is an open set, by theorem 21, $\{t \in V | \Gamma(t) \cap \Theta\}$ is open. That entails, from remark 2 of n°12, that Γ is continuous on V for the Hausdorff topology.

23 - Corollary II-23. Let T be a topological space, and Γ map T to convex compact non empty subsets of $\mathbb{R}^n$. Then if the support functions $\delta^*(x'|\Gamma(.))$ are continuous, Γ is continuous.

Proof. Let $(e'_1, \ldots, e'_n)$ be a basis of $(\mathbb{R}^n)'$. Then if $t_0 \in T$ there exists a neighbourhood of t_0, V, such that the functions $\delta^*(e'_i|\Gamma(.))$ and $\delta^*(-e'_i|\Gamma(.))$ $(i \in \{1,\ldots,n\})$ are bounded on V. Hence $\bigcup_{t \in V} \Gamma(t)$ is bounded. Thus we may apply corollary 22.

24 - Corollary II-24. Let T be a topological space which is locally compact or metrizable, E a Hausdorff locally convex space, and Γ map T to convex weakly compact non empty subsets of E. Then if the support functions $\delta^*(x'|\Gamma(.))$ are continuous, Γ is continuous for the Hausdorff topology corresponding to $\sigma(E,E')$. If moreover E is Montel, Γ is continuous for the Hausdorff topology.

Proof. 1) By theorem 20 Γ is u.s.c. for the weak topology. If T is locally compact we may suppose T is compact. If T is metrizable it is sufficient to prove continuity properties on sets such as $\{t_n, \bar{t} | n \in N\}$

where $t_n \to \bar{t}$ (because Γ l.s.c. at $\bar{t}$ (respectively u.s.c.) $\Leftrightarrow \forall (t_n)$ which converge to $\bar{t}$ and $\forall$ open U, if $\Gamma(\bar{t}) \cap U \neq \emptyset$ (resp. $\Gamma(\bar{t}) \subset U$), then for n large enough $\Gamma(t_n) \cap U \neq \emptyset$ (resp. $\Gamma(t_n) \subset U$)).

Hence we may always suppose T is compact. By a theorem of Berge (theorem 25 below) $\bigcup_{t \in T} \Gamma(t)$ is weakly compact. Then by theorem 21, Γ is l.s.c. Finally if E is enlowed with the weak Hausdorff uniformity, Γ is continuous thanks to remark 2 of n°12.

2) If E is a Montel space, the set $\bigcup \Gamma(t)$ in the first part is compact for the strong topology of E. Hence Γ is u.s.c. for the strong topology (because if U is open, $U \cap (\bigcup \Gamma(t))$ is also a weakly open set). And Γ is l.s.c. for the strong topology by theorem 21.

25 - <u>Theorem II-25</u>. (Berge) - <u>Let</u> T <u>be a compact space</u>, E <u>a Hausdorff space, and</u> Γ <u>a multifunction from</u> T <u>to compact subsets of</u> E. <u>Then if</u> Γ <u>is u.s.c. th set</u> $\bigcup_{t \in T} \Gamma(t)$ <u>is compact.</u>

<u>Proof</u>. Let $(U_i)_{i \in I}$ a family of open sets which covers $\bigcup_{t \in T} \Gamma(t)$.

Each $\Gamma(t)$ is covered by an open set, V_t, which is the union of a finite subfamily of (U_i). The set

$$T_\theta = \{t | \Gamma(t) \subset V_\theta\} \quad \text{contains } \theta \text{ and is open.}$$

Hence $(T_\theta)_{\theta \in T}$ covers T. As T is compact, there exist $\theta_1, \ldots, \theta_n$ such that $T = T_{\theta_1} \cup \ldots \cup T_{\theta_n}$.

Then $\bigcup_{t \in T} \Gamma(t)$ is contained in $V_{\theta_1} \cup \ldots \cup V_{\theta_n}$ which is the union of a finite subfamily of $(U_i)_{i \in I}$.

BIBLIOGRAPHY OF CHAPTER II

1 - BERGE, C. - Espaces topologiques. Fonctions multivoques. Dunod (1959).

2 - BOURBAKI, N. - Topologie générale. Hermann.

There are a lot of exercises about topology on spaces of subsets : Ch. I : § 2, ex. 7 ; § 3, ex. 10, 16 ; § 5, ex. 7 ; § 8, ex. 12, 29 ; § 9, ex. 13, 14 ; § 10, ex. 20 ; § 11, ex. 18, 24 ; - Ch. II : § 1, ex. 5 ; § 2, ex. 6 ; § 3, ex. 7 ; § 4, ex. 11, 15 ; - Ch. IX : § 2, ex. 6, 21 ; - Ch. X : § 1, ex. 7 -

See also Espaces vectoriels topologiques - Ch. II, § 2, ex. 39.

3 - CHOQUET, G. - Convergences. Ann. Univ. Grenoble 23 (1947), 55-112.

4 - DEBREU, G. - Integration of correspondences. 5th Berkeley Symposium on Math. Stat. Prob. Vol. II, part. I, p. 351 - 372.

5 - GODET-THOBIE, C. - PHAM THE LAI - Sur le plongement de l'ensemble des convexes fermés bornés d'un espace vectoriel topologique localement convexe dans un espace vectoriel topologique localement convexe.

Comptes Rendus Ac. Sc. Paris 271 (1970), 84 - 87.

6 - HORMANDER, L. - Sur la fonction d'appui des ensembles convexes dans un espace localement convexe.

Arkiv. för Math. 3-12 (1954), 181-186.

7 - JOLY, J.L. - Une famille de topologies et de convergences sur l'ensemble des fonctionnelles convexes - Thèse Grenoble 1970.

8 - KURATOWSKI, K. - Topology vol. I - Academic Press (1966).

9 - MICHAEL, E. - Topologies on spaces of subsets. Tr. A.M.S. 71 (1951), 152-182.

10 - MOREAU, J.J. - Fonctionnelles convexes. Séminaire sur les équations aux dérivées partielles II. Collège de France 1966-67.

11 - Rétraction d'une multi-application. Séminaire d'Analyse Convexe Montpellier 1972, exposé n°13.

12 - PALLU DE LA BARRIERE, R. Quelques propriétés des multimesures. Séminaire d'Analyse Convexe. Montpellier 1973 exposé n°11.

13 - RADSTROM, H. - An embedding theorem for spaces of convex sets. Proc. A.M.S. 3 (1952) - 165-169.

14 - ROCKAFELLAR, R.T. - Convex Analysis - Princeton University Press (1970).

15 - SAINT RAYMOND J. - Topologie sur l'ensemble des compacts non vides d'un espace topologique séparé. Séminaire Choquet 1969/70 n°21.

16 - VALADIER, M. - Contribution à l'Analyse Convexe. Thèse, Paris (1970).

17 - VAN CUTSEM, B. - Eléments aléatoires à valeurs convexes compactes. Thèse Grenoble (1971).

CHAPTER III

MEASURABLE MULTIFUNCTIONS

Consider a measurable space $(T,\mathcal{C})$, a complete separable metric space X and Γ a multifunction from T to non empty closed subsets of X. An important problem is the existence of a measurable selection of Γ : a selection is a map $\sigma : T \to X$ such that $\forall t,\ \sigma(t) \in \Gamma(t)$. Denote by $\Gamma^-(B)$ the set $\{t \mid \Gamma(t) \cap B \neq \emptyset\}$ and by G the graph of Γ : $\{(t,x) \in T\times X \mid x \in \Gamma(t)\}$. Consider the following properties :

(i) $\forall B$ borel, $\Gamma^-(B) \in \mathcal{C}$,

(ii) $\forall F$ closed, $\Gamma^-(F) \in \mathcal{C}$,

(iii) $\forall U$ open, $\Gamma^-(U) \in \mathcal{C}$,

(iv) there exists a sequence (σ_n) of measurable selections such that $\forall t,\ \Gamma(t) = \overline{\{\sigma_n(t) \mid n \in \mathbb{N}\}}$

(v) $\forall x \in X$, $d(x,\Gamma(.))$ is measurable

(vi) the graph of Γ belongs to $\mathcal{C} \otimes \mathcal{B}(X)$.

We shall prove that :

(i) $\Rightarrow$ (ii) $\Rightarrow$ (iii) $\Leftrightarrow$ (iv) $\Leftrightarrow$ (v) $\Rightarrow$ (vi) (§2 th.9 prop.11 and 13)

It is natural to say that Γ is measurable if (iii) (or (iv) or (v)) is verified. When Γ is compact valued the measurability of Γ is equivalent to the measurability of Γ considered as a map from T to the set of all compact subsets of X endowed with the Hausdorff metric (§1). When X is a Fréchet space and Γ is convex compact valued, then the measurability of Γ is equivalent to "$\forall x' \in X'$, $\delta^*(x' \mid \Gamma(.))$ is measurable" (§3).

One may ask : are the six properties (i)...(vi) equivalent? This is true when $\mathcal{C}$ is a complete tribe : indeed by the projection theorem (vi) $\Rightarrow$ (i) (§4).

For some applications it is interesting to obtain measurable selections with X Lusin or Suslin, and Γ non closed valued. That is the von Neumann-Aumann theorem which says that if $\mathcal{T}$ is complete, Γ is non empty valued and the graph of Γ is measurable then Γ admits measurable selections (§4).

Finally some measurable convex valued multi-functions are considered in §5 and some implicit measurable function theorems are given in §6.

§0 - PREREQUISITES

A measurable space $(T,\mathcal{T})$ is a pair where T is a set and $\mathcal{T}$ is a tribe (or σ-field) of subsets of T. Recall that $\mathcal{T}$ is a tribe if

- $\emptyset \in \mathcal{T}$,
- $A \in \mathcal{T} \Rightarrow T-A \in \mathcal{T}$,
- $\forall n \in \mathbb{N},\ A_n \in \mathcal{T} \Rightarrow \cup A_n \in \mathcal{T}$.

The elements of $\mathcal{T}$ are called measurable sets.

If U is a topological space, the Borel tribe $\mathcal{B}(U)$ is the smallest tribe containing all the open sets. If $(T,\mathcal{T})$ and $(U,\mathcal{U})$ are two measurable spaces, the product tribe $\mathcal{T} \otimes \mathcal{U}$ on $T\times U$ is the smallest tribe containing all the $A\times B$ $(A \in \mathcal{T},\ B \in \mathcal{U})$.

If $(T,\mathcal{T})$, $(U,\mathcal{U})$ are two measurable spaces, a function $f:T \to U$ is said to be $(\mathcal{T},\mathcal{U})$-measurable if $\forall\ B \in \mathcal{U},\ f^{-1}(B) \in \mathcal{T}$. This property enables us to carry a positive or bounded measure on T into a measure on U). If U is a topological space, a $(\mathcal{T},\mathcal{B}(U))$ measurable function is called a Borel function. When T and U are both topological spaces, a continuous function is Borel.

If $(T,\mathcal{T})$ is a measurable space and U a metric space, we say that $f:T \to U$ is (strongly) measurable if one of the following equivalent properties is satisfied :

(i) f is $(\mathcal{T},\mathcal{B}(U))$-measurable and $f(T)$ is separable,

(ii) f is the pointwise limit of a sequence of measurable functions assuming a finite number of values,

(iii) f is the uniform limit of a sequence of measurable functions assuming a countable number of values.

This property ensures that if U is a topological group, ological group, and if f and g are measurable, then $f+g$ is measurable).

If U is a metric space, if f_n is a sequence of $(\mathcal{T},\mathcal{B}(U))$ (resp. strongly) measurable functions from T to U, and if $f_n \to f$ pointwise, then f is $(\mathcal{T},\mathcal{B}(U))$ (resp. strongly) measurable.

If $(T,\mathcal{T})$ is a measurable space, a bounded (resp. positive) measure on $(T,\mathcal{T})$ is a map $\mu: \mathcal{T} \to \mathbb{R}$ (resp. $\mu: \mathcal{T} \to [0,\infty]$) such that for countable pairwise disjoint A_n belonging to $\mathcal{T}$, $\mu(\cup A_n) = \Sigma\, \mu(A_n)$. A positive measure is σ-finite if T is the union of a sequence of measurable sets of finite measure.

If μ is a positive measure on $(T,\mathcal{T})$ one says that a subset N of T is negligible if there exists $A \in \mathcal{T}$ such that $N \subset A$ and $\mu(A) = 0$. The μ-completion of $\mathcal{T}$, is the tribe generated by $\mathcal{T}$ and the negligible sets ; it is denoted by $\mathcal{T}_\mu$ The measure μ admits a unique extension to $\mathcal{T}_\mu$. The tribe $\mathcal{T}$ is said to be μ-complete if $\mathcal{T} = \mathcal{T}_\mu$.

If μ is a positive measure on $(T,\mathcal{T})$, U a metric space, and $f:T \to U$, we say that f is measurable (or more precisely μ-measurable) if for every $A \in \mathcal{T}$ of finite measure, there exists a negligible set N such that the restriction $f_{|A-N}$ is strongly measurable.

Let T be a Hausdorff topological space. A positive <u>Radon measure</u> μ on T is a positive measure $\mu : \mathcal{B}(T) \to [0,\infty]$ such that

- $\forall t \in T$, there exists an open neighbourhood of t of finite measure,

- $\forall A \in \mathcal{B}(T)$, $\mu(A) = \sup \{\mu(K) \mid K \text{ compact}, K \subset A\}$.

Let T be a Hausdorff topological space, μ a positive Radon measure on T, U a topological space, and $f : T \to U$. We say that f is <u>Lusin μ-measurable</u> if :

$\forall K$ compact, $K \subset T$, $\forall \epsilon > 0$, $\exists L$ compact, $L \subset K$ such that $\mu(K-L) < \epsilon$ and $f_{|L}$ is continuous.

If moreover U is a metric space then,

f is Lusin μ-measurable $\Leftrightarrow$ $\forall K$ compact, $f_{|K}$ is μ-measurable.

$\Leftrightarrow$ f is μ-measurable.

§1 - MEASURABLE MULTIFUNCTIONS WITH VALUES IN COMPACT SUBSETS OF A METRIZABLE SEPARABLE SPACE.

1 - Let $(T,\mathcal{T})$ be a measurable space. Let X be a metrizable separable space.

<u>Definition. A multifunction Γ from T to compact subsets of X is said to be measurable if it is measurable as a function from T to $\mathcal{P}_k(X)$ (with the Hausdorff topology defined in ch. II).</u>

2 - <u>Theorem III.2. With the hypotheses of definition 1, Γ is measurable is equivalent to any of the following properties.</u>

a) $\forall U$ <u>open set in</u> X, $\Gamma^-(U) = \{t \in T \mid \Gamma(t) \cap U \neq \emptyset\} \in \mathcal{T}$

b) $\forall F$ <u>closed set in</u> X, $\Gamma^-(F) = \{t \in T \mid \Gamma(t) \cap F \neq \emptyset\} \in \mathcal{T}$

Proof. We shall apply theorem II-10.

1) Remark that

$\Gamma^{-}(U) = \Gamma^{-1}(\{K \in \mathcal{P}_k(X) | K \cap U \neq \emptyset\})$. By theorem II-10 $\{K | K \cap U \neq \emptyset\}$ is a borel set (it is even open by theorem II-6). Hence if Γ is measurable a) is true. The converse is true by theorem II-10 : the sets $\{K | K \cap U \neq \emptyset\}$ generate the Borel tribe of $\mathcal{P}_k(X)$.

2) To prove "Γ measurable" $\Leftrightarrow$ b, it suffices to remark that $\Gamma^{-}(F) = \complement_T \Gamma^{-1}(\{K | K \subset X - F\})$ and to apply again theorem II-10.

3 - Corollary III.3. *If T is a topological space, and Γ is a multifunction from T to $\mathcal{P}_k(X)$, if Γ is upper semi-continuous (or lower semi-continuous) then Γ is measurable (with respect to the Borel tribe $\mathcal{B}(T)$)* (definitions of u.s.c. and l.s.c. have been given in II.20. and II.21.).

Proof. If Γ is u.s.c. then for every closed set F, $\{t | \Gamma(t) \subset X-F\}$ is open, hence $\Gamma^{-}(F)$ belongs to $\mathcal{B}(T)$. If Γ is l.s.c. then for every open set U, $\{t | \Gamma(t) \cap U \neq \emptyset\}$ is open, hence it belongs to $\mathcal{B}(T)$.

Remark. If T is a Hausdorff topological space and μ a positive Radon measure on T, and if Γ satisfies definition 1, then for each integrable set $T_0 \subset T$, and each $\varepsilon > 0$, there exists a compact $T_1 \subset T_0$ such that $\mu(T_0 - T_1) < \varepsilon$ and Γ is continuous on T_1. That is the Lusin property.

4 - Proposition III.4. *If Γ_1 and Γ_2 are compact valued measurable multifunctions then the multifunction $t \mapsto \Gamma_1(t) \cap \Gamma_2(t)$ is measurable. If (Γ_n) is a sequence of compact valued measurable multifunctions then $t \mapsto \cap \Gamma_n(t)$ is measurable, and, if $\overline{\cup \Gamma_n(t)}$ is compact, $t \mapsto \overline{\cup \Gamma_n(t)}$ is measurable.*

Proof.

1) Let us prove that the map $(K_1, K_2) \mapsto K_1 \cap K_2$ from $(\mathcal{S}_k(X))^2$ to $\mathcal{S}_k(X)$ is Borel. Let U be an open set in X. Let us prove that $\{(K_1, K_2) | K_1 \cap K_2 \subset U\}$ is open. Indeed if $K_1^0 \cap K_2^0 \subset U$, $K_1^0 - U$ and $K_2^0 - U$ are disjoint. Then there exist two open sets U_1 and U_2 such that $K_i^0 - U \subset U_i$ $(i = 1,2)$ and $U_1 \cap U_2 = \emptyset$ (the important fact is that $K_i^0 - U$ is closed and X is normal. See Kuratowski). By theorem II-6 $\{(K_1, K_2) | K_1 \subset U_1 \cup U, K_2 \subset U_2 \cup U\}$ is a neighbourhood of (K_1^0, K_2^0). For such a (K_1, K_2) one has

$$K_1 \cap K_2 \subset (U_1 \cup U) \cap (U_2 \cup U) = U .$$

Therefore, by theorem II-10, $(K_1, K_2) \mapsto K_1 \cap K_2$ is borel. Hence $t \mapsto \Gamma_1(t) \cap \Gamma_2(t)$ is measurable.

2) Applying the first part, we obtain that $\Gamma'_n = \Gamma_0 \cap \ldots \cap \Gamma_n$ is measurable. But $\Gamma'_n(t)$ converges to $\cap \Gamma_n(t)$ for the Hausdorff distance (it is easy to see that $e(\Gamma'_n(t), \cap \Gamma_n(t)) \to 0$). Hence $t \mapsto \cap \Gamma_n(t)$ is measurable.

3) Let U be an open set of X. Then

$$\begin{aligned} &\{t | \overline{\cup \Gamma_n(t)} \cap U \neq \emptyset\} \\ &= \{t | \cup \Gamma_n(t) \cap U \neq \emptyset\} \\ &= \bigcup_n \{t | \Gamma_n(t) \cap U \neq \emptyset\} \text{ belongs to } \mathcal{C}. \end{aligned}$$

Hence, by theorem 2, $\overline{\cup \Gamma_n(.)}$ is measurable.

§2 - SELECTION THEOREM. MEASURABLE MULTIFUNCTIONS WITH VALUES IN COMPLETE SUBSETS OF A SEPARABLE METRIC SPACE.

5 - Definition. Let $\Gamma : T \to \mathcal{S}(X)$. A function $\sigma : T \to X$ will be said to be a selection of Γ if $\sigma(t) \in \Gamma(t)$ for every t.

An important problem is, when Γ is non empty valued and has some property of measurability, to prove existence of measurable selections. The fundamental theorem is theorem 6, below, and its consequences, theorems 7 and 8. But a very general and useful theorem is theorem 22.

6 - Theorem III.6. Let X be a separable metric space, $(T, \mathcal{C})$ a measurable space, Γ a multifunction from T to complete non empty subsets of X. If for each open set U in X, $\Gamma^{-}(U)$ $(= \{t \mid \Gamma(t) \cap U \neq \emptyset\})$ belongs to $\mathcal{C}$, then Γ admits a measurable selection.

Proof. Let $\{x_n\}$ be a countable dense set in X. We define a sequence of measurable functions assuming a countable number of values, (σ_p) by recurrence, with the properties $d(\sigma_p(t), \Gamma(t)) < 2^{-p}$, $d(\sigma_{p+1}(t), \sigma_p(t)) \leq 2^{-p+1}$

First we put $\sigma_0(t) = x_n$ if n is the smallest integer such that $\Gamma(t) \cap B(x_n, 2^0) \neq \emptyset$ ($B(x_n, r)$ is the open ball of radius r and center x_n). Thus σ_0 is measurable :

$$\sigma_0^{-1}(x_n) = \Gamma^{-}(B(x_n, 2^0)) - \bigcup_{m<n} \Gamma^{-}(B(x_m, 2^0)).$$

Suppose now σ_p is chosen. Let $T_i = \sigma_p^{-1}(x_i)$. Then if $t \in T_i$, $\Gamma(t) \cap B(x_i, 2^{-p}) \neq \emptyset$. We put, on T_i, $\sigma_{p+1}(t) = x_n$ if n is the smallest integer such that $\Gamma(t) \cap B(x_i, 2^{-p}) \cap B(x_n, 2^{-(p+1)}) \neq \emptyset$. Hence σ_{p+1} is measurable, $d(\sigma_{p+1}(t), \Gamma(t)) < 2^{-(p+1)}$, and $d(\sigma_{p+1}(t), \sigma_p(t)) \leq 2^{-p} + 2^{-(p+1)} \leq 2^{-p+1}$. From the last inequality follows that $(\sigma_p(t))$ is a Cauchy sequence. As $\Gamma(t)$ is complete and $d(\sigma_p(t), \Gamma(t)) \to 0$, the limit of $(\sigma_p(t))$ in the completion of X belongs to $\Gamma(t)$. This limit $\sigma(t)$ defines a measurable selection of Γ.

7 - Theorem III.7. Under the same hypothesis as in theorem 6, there exists a sequence (σ_n) of measurable selections of Γ such that for every t, $\Gamma(t) = \overline{\{\sigma_n(t) | n \in \mathbb{N}\}}$.

Proof. Let $\{x_n\}$ be a dense set in X. For $(n,i) \in \mathbb{N}^2$ set

$$\Gamma_{ni}(t) = \begin{cases} \Gamma(t) \cap B(x_n, 2^{-i}) & \text{if } t \in \Gamma^-(B(x_n, 2^{-i})) \\ \Gamma(t) & \text{otherwise.} \end{cases}$$

The multifunction $t \mapsto \overline{\Gamma_{ni}(t)}$ has non empty complete values. For any open set U,

$$\{t | \overline{\Gamma_{ni}(t)} \cap U \neq \emptyset\} = \{t | \Gamma_{ni}(t) \cap U \neq \emptyset\}$$
$$= \Gamma^-(B(x_n, 2^{-i}) \cap U) \cup [\complement \Gamma^-(B(x_n, 2^{-i})) \cap \Gamma^-(U)] \in \mathcal{C}.$$

Hence, by theorem 6, $\overline{\Gamma_{ni}}$ has a measurable selection σ_{ni}.

Now let us show that $\Gamma(t) = \overline{\{\sigma_{ni}(t)\}}$. Let $x \in \Gamma(t)$ and $\varepsilon > 0$. Let i be such that $2^{-i} \le \frac{\varepsilon}{2}$, and n such that $d(x_n, x) < 2^{-i}$. Hence $t \in \Gamma^-(B(x_n, 2^{-i}))$ and $\sigma_{ni}(t) \in B(x_n, 2^{-i})$.

Hence $d(\sigma_{ni}(t), x) \le d(\sigma_{ni}(t), x_n) + d(x_n, x) \le \varepsilon$.

8 - Theorem III.8. 1) Let $(T, \mathcal{C})$ be a measurable space, X a Polish space, and Γ map T to non empty closed subsets of X. If for every open set U in X, $\Gamma^-(U) \in \mathcal{C}$, then Γ admits a sequence of measurable selections (σ_n) such that $\Gamma(t) = \overline{\{\sigma_n(t)\}}$.

2) *Let* $(T,\mathcal{T})$ *be a measurable space,* X *a separable metric space, and* Γ *map* T *to non empty compact subsets of* X. *If* Γ *is measurable* (*see definition* 1), *then* Γ *admits a sequence of measurable selections* (σ_n) *such that* $\Gamma(t) = \overline{\{\sigma_n(t)\}}$.

Proof. It is an easy consequence of theorem 7.

9 - *Theorem III.9.* *Let* $(T,\mathcal{T})$ *be a measurable space,* X *a separable metric space, and* Γ *map* T *to non empty complete subsets of* X. *Then the following properties are equivalent*

a) $\Gamma^-(U) \in \mathcal{T}$ *for every open set* U,

b) $d(x, \Gamma(.))$ *is measurable for every* $x \in X$,

c) Γ *admits a sequence of measurable selections* (σ_n) *such that* $\Gamma(t) = \overline{\{\sigma_n(t)\}}$.

Proof. $a \Rightarrow c$ is theorem 7.

$c \Rightarrow b$ because

$$d(x, \Gamma(t)) = \inf \{d(x, \sigma_n(t)) | n \in N\} .$$

$b \Rightarrow a$ Remark that $\{t | d(x,\Gamma(t)) < r\} = \Gamma^-(B(x,r))$ ($r > 0$, $B(x,r)$ denotes the open ball). But any open set U is the union of a sequence of balls $B(x_n, r_n)$. Hence $\Gamma^-(U) = \bigcup_n \Gamma^-(B(x_n, r_n))$ is measurable if b is true.

Remark. It is easy to prove $a \Rightarrow b$ and $c \Rightarrow a$:

$a \Rightarrow b$ by the formula $\{t | d(x, \Gamma(t)) < r\} = \Gamma^-(B(x, r))$ ($r > 0$)

and $c \Rightarrow a$ by the formula

$$\Gamma^-(U) = \bigcup \sigma_n^{-1}(U).$$

10 - Definition. *If* $(T,\mathcal{T})$ *is a measurable space,* X *a separable metric space, and* Γ *map* T *to complete subsets of* X, *then* Γ *will be said to be measurable if* $T_0 = \{t \mid \Gamma(t) = \emptyset\}$ *belongs to* $\mathcal{T}$ *and if on* $T-T_0$, Γ *has the properties of theorem 9.*

Remark. By theorem 8 2) the definitions 1 and 10 are consistent.

There are two other properties of measurability that one could take as definition of measurability. They are :

"$\Gamma^-(F) \in \mathcal{T}$ for every closed set F" and

"the graph of Γ (that is $\{(t,x) \in T \times X \mid x \in \Gamma(t)\}$) belongs to $\mathcal{T} \otimes \mathcal{B}(X)$".

We examine these properties in the three following propositions. For some σ-field $\mathcal{T}$, all five properties are equivalent (see theorem 30 below).

11 - Proposition III.11. *Let* $(T,\mathcal{T})$ *be a measurable space,* X *a metric space, and* Γ *map* T *to* $\mathcal{S}(X)$. *Then if* $\Gamma^-(F) \in \mathcal{T}$ *forevery closed set* F, *then* $\Gamma^-(U) \in \mathcal{T}$ *for every open set* U.

Proof. Every set U is F_σ : it suffices to put
$F_n = \{x \in X \mid d(x,X-U) \geq \frac{1}{n}\}$ $(n \geq 1)$. Then $U = \cup F_n$, and $\Gamma^-(U) = \cup \Gamma^-(F_n)$.

12 - Proposition III.12. 1) *Let* $(T,\mathcal{T})$ *be a measurable space,* X *a locally compact Polish space, and* Γ *map* T *to* $\mathcal{S}_f(X)$. *If* $\Gamma^-(U) \in \mathcal{T}$ *for every open set* U, *then* $\Gamma^-(F) \in \mathcal{T}$ *for every closed set* F.

2) *Let* $(T,\mathcal{T})$ *be a measurable space,* X *a metrizable space, and* Γ *map* T *to* $\mathcal{S}_k(X)$. *If* $\Gamma^-(U) \in \mathcal{T}$ *for every open set* U, *then,* $\Gamma^-(F) \in \mathcal{T}$ *for every closed set* F.

Proof. 1) If F is closed, F is the union of a sequence of compact sets K_n. Then $\Gamma^-(F) = \cup \Gamma^-(K_n)$. Suppose K is compact and let us prove $\Gamma^-(K) \in \mathcal{C}$. The Alexandroff compactified $\hat{X}$ of X is metrizable (that is equivalent for a locally compact space to say it is Polish). Let d be a metric on $\hat{X}$. Then for n large enough $(n \geq n_o)$ $K_n = \{x \in \hat{X} | d(x,K) \leq \frac{1}{n}\}$ is contained in X, and compact.

Let (σ_n) be a sequence of selections of Γ as in theorem 7.

Then $\Gamma^-(K) = \bigcap_{n \geq n_o} \bigcup_m \sigma_m^{-1}(K_n)$. Indeed if $t \in \Gamma^-(K)$, $\Gamma(t) \cap K \neq \phi$ implies for every $n \geq n_o$, $\Gamma(t) \cap K_n \neq \phi$ and therefore there exists m such that $\sigma_m(t) \in K_n$. Conversely, if $t \in \bigcap_n \bigcup_m \sigma_m^{-1}(K_n)$, then for every n, $\Gamma(t) \cap K_n \neq \phi$. Let $x_n \in \Gamma(t) \cap K_n$. The sequence $(x_n)_{n \geq n_o}$ is contained in K_{n_o}. Let $\bar{x}$ be a cluster point of (x_n). Then $\bar{x} \in \cap K_n = K$ and $\bar{x} \in \Gamma(t)$.

2) If F is closed, put $U_n = \{x | d(x,F) < \frac{1}{n}\}$. Then $\Gamma^-(F) = \cap \Gamma^-(U_n)$ because if $t \in \cap \Gamma^-(U_n)$, let $x_n \in \Gamma(t) \cap U_n$. As $\Gamma(t)$ is compact, the sequence (x_n) has a cluster point $\bar{x}$. And $\bar{x} \in \Gamma(t) \cap F$. (it is the same argument as in the second part of the proof of II.10).

13 - Proposition III.13. If $(T, \mathcal{C})$ is a measurable space, X a separable metric space, and if Γ from T to complete subsets of X is measurable (see definition 10), then the graph of Γ (that is $\{(t,x) \in T \times T | x \in \Gamma(t)\}$) belongs to $\mathcal{C} \otimes \mathcal{B}(X)$.

Proof. We may suppose $\Gamma(t)$ non empty forevery t. The graph G of Γ is $G = \{(t,x) | d(x, \Gamma(t)) = 0\}$. But the function $d(x, \Gamma(.))$ is measurable (that is the property of measurability of Γ which is used. Completeness of $\Gamma(t)$ is not necessary !), so, by the following lemma $(t, x) \mapsto d(x, \Gamma(t))$ is measurable, and $G \in \mathcal{C} \otimes \mathcal{B}(X)$.

14 - Lemma III.14. Let $(T, \mathcal{C})$ be a measurable space, X a separable metrizable space, U a metrizable space and $\varphi : T \times X \to U$. We suppose that φ is measurable (resp. $(\mathcal{C}, \mathcal{B}(U))$ measurable) in t and continuous in x. Then φ is measurable (resp. $(\mathcal{C} \otimes \mathcal{B}(X), \mathcal{B}(U))$ measurable).

Proof. Let (x_n) be a dense sequence in X. For $p \geq 1$ put $\varphi_p(t,x) = \varphi(t,x_n)$ if n is the smallest integer such that x belongs to $B(x_n, \frac{1}{p})$ (the open ball with center x_n and radius $\frac{1}{p}$). It is clear that $\varphi_p(t,x) \to \varphi(t,x)$ as $p \to \infty$. And φ_p is measurable (resp. $(\mathcal{C} \otimes \mathcal{B}(X), \mathcal{B}(U))$ measurable) because on

$$T \times [B(x_n, \frac{1}{p}) - \bigcup_{m<n} B(x_m, \frac{1}{p})],$$

φ_p is equal to the function $(t,x) \mapsto \varphi(t, x_n)$.

§3 - MEASURABLE COMPACT CONVEX MULTIFUNCTIONS

15 - Theorem III.15. Let $(T, \mathcal{C})$ be a measurable space, E a locally convex metrizable separable vector space, and Γ map T to $\mathcal{P}_{ck}(E)$ (the space of convex compact subsets of X). Then Γ is measurable if and only if the support functions $\delta^*(x' | \Gamma(.))$ are measurable.

Proof. If Γ is measurable, $T_0 = \{t | \Gamma(t) = \emptyset\} \in \mathcal{C}$. If $\delta^*(x' | \Gamma(.))$ is measurable then $T_0 = \{t | \delta^*(0 | \Gamma(t)) = -\infty\} \in \mathcal{C}$. Hence we may suppose $\Gamma(t)$ non empty. If Γ is measurable, by theorem 7 there exists a sequence of measurable selections (σ_n), such that $\Gamma(t) = \overline{\{\sigma_n(t)\}}$. So $\delta^*(x'|\Gamma(t)) = \sup_n \langle x', \sigma_n(t)\rangle$ is a measurable function of t. Conversely suppose $\delta^*(x'|\Gamma(.))$ measurable. There exists an increasing sequence of semi-norms (p_n) which define the topology of E. Let h_n be the Hausdorff semi-distance associated with p_n. By theorem II.12 the distance on $\mathcal{P}_{ck}(E)$, $H(A,B) = \Sigma\, 2^{-n} \dfrac{h_n(A,B)}{1+h_n(A,B)}$ defines the uniform structure of $\mathcal{P}_{ck}(E)$. We shall prove the following : for every $A \in \mathcal{P}_{ck}(E)$, $t \mapsto H(A,\Gamma(t))$ is measurable. That, by lemma 16 below (applied to Γ in place of f and $\mathcal{P}_{ck}(E)$ in place of Z (see theorem II.8)), entails the theorem. It is sufficient to prove measurability of $h_n(A, \Gamma(.))$. But by theorem II.18, $h_n(A, B) = \sup \{|\delta^*(x'|A) - \delta^*(x'|B)| \mid x' \in V_n^o\}$ (with $V_n = \{x | p_n(x) \leq 1\}$). The set V_n^o is equicontinuous, hence compact for the topology of uniform convergence on compact subsets of E. Let $\{x_n'\}$ be a countable dense set in V_n^o. As $\delta^*(.|A)$ and $\delta^*(.|\Gamma(t))$ are continuous for the above topology, we have

$$h_n(A, \Gamma(t)) = \sup_n |\delta^*(x_n'|A) - \delta^*(x_n'|\Gamma(t))| .$$

Hence it is measurable in t.

Remark. When E is a normed space, it suffices to prove that $\forall x \in E$, $d(x, \Gamma(.))$ is measurable (thanks to theorem 9).

16 - Lemma III.16. Let $(T, \mathcal{T})$ be a measurable space, Z a separable metric space and $f : T \to Z$, such that for every $z \in Z$, $d(z, f(.))$ is measurable. Then f is measurable.

Sophisticated proof.

The set-valued function $t \mapsto \{f(t)\}$ has a measurable selection by theorem 9.

§4 - PROJECTION THEOREM. VON NEUMANN-AUMANN SELECTION THEOREM

17 - Definition. A Suslin space is a Hausdorff topological space such that there exists a Polish space P and a continuous map from P onto S.

For properties of Suslin spaces see Bourbaki [2].

18 - In the following we shall denote by pr_T the projection map from a product $T \times U$ onto T.

Theorem III.18. Let T be a Suslin space, X a topological space, and Γ map T to non empty subsets of X, whose graph G is Suslin. Denote by $\mathcal{T}_s$ the σ-field on T generated by Suslin subsets of T. Then there exists a sequence (σ_n) of selections of Γ, which are $(\mathcal{T}_s, \mathcal{B}(X))$ measurable and such that for every $t \in T$, $\{\sigma_n(t)\}$ is dense in $\Gamma(t)$. Moreover if P is Polish and $h : P \to G$ is continuous and onto, the σ_n can be chosen such that $\sigma_n = pr_X \circ h \circ \rho_n$ where $\rho_n : T \to P$ is $\mathcal{T}_s$ measurable. Consequently every σ_n is the limit of a sequence of $\mathcal{T}_s$ measurable functions assuming a finite number of values, and if in addition μ is a Radon measure on T, σ_n is Lusin μ-measurable.

Proof. Let P be Polish and $h : P \to G$ continuous and onto. Let $\pi : G \to T$ be the map $(t,x) \mapsto t$. The set-valued function $\Sigma = (\pi \circ h)^{-1}$ has a closed graph H in $T \times P$.

Therefore if U is open in P, $\Sigma^-(U) = pr_T(H \cap T \times U)$ is Suslin. By theorem 8 there exists a sequence (ρ_n) of $\mathcal{T}_s$ measurable selections of Σ, such that for every t, $\{\rho_n(t)\}$ is dense in $\Sigma(t)$. From $\rho_n(t) \in (\pi \circ h)^{-1}(t)$ results $t = \pi \circ h \circ \rho_n(t)$, hence $h \circ \rho_n(t) \in \{t\} \times \Gamma(t)$. Put $h \circ \rho_n(t) = (t, \sigma_n(t))$. Then σ_n is a selection of Γ and the formula $\sigma_n = pr_X \circ h \circ \rho_n$ proves the measurability properties. Finally, the map $h : \Sigma(t) \to \{t\} \times \Gamma(t)$ is continuous and onto, so the $h \circ \rho_n(t)$ are dense in $\{t\} \times \Gamma(t)$ and the $\sigma_n(t)$ are dense in $\Gamma(t)$.

19 - The following theorem is due in a primary form to Von Neumann. [28].

Theorem III.19. Let S_1 and S_2 be Suslin spaces and $\varphi : S_1 \to S_2$ a continuous map onto S_2. Then there exists a map $\sigma : S_2 \to S_1$ such that $\varphi \circ \sigma(s) = s$ for every $s \in S_2$, and σ is measurable for the σ-field $\mathcal{C}_s(S_2)$ generated by Suslin subsets of S_2 and the Borel σ-field on S_1. Moreover σ is also the limit of a sequence of $\mathcal{C}_s(S_2)$ measurable functions assuming a finite number of values, and if in addition μ is a Radon measure on S_2, σ in Lusin μ-measurable.

Proof. This follows from theorem 18 with $T = S_2$, $X = S_1$ and $\Gamma = \varphi^{-1}$. Indeed the graph of φ^{-1} is closed (because φ is continuous) in $S_2 \times S_1$, hence Suslin.

20 - Lemma III.20. Let X be a topological space and $Y \subset X$. Then $\mathcal{B}(Y) = \mathcal{B}(X) \cap Y$ (where $\mathcal{B}(X) \cap Y$ denotes

$$\{B \cap Y \mid B \in \mathcal{B}(X)\}).$$

Proof. The embedding map $i : Y \to X$ is continuous, hence Borel measurable, and if $B \in \mathcal{B}(X)$, $i^{-1}(B) = B \cap Y$ belongs to $\mathcal{B}(Y)$. That proves $\mathcal{B}(X) \cap Y \subset \mathcal{B}(Y)$. Conversely $\mathcal{B}(Y)$ is generated by the open sets of Y. But any open set U in Y is the intersection of Y and an open set of X, that is $U \in \mathcal{B}(X) \cap Y$.

21 - Definition. Let $(T, \mathcal{C})$ be a measurable space. If μ is a positive measure on $(T, \mathcal{C})$, $\mathcal{C}_\mu$ denotes the μ-completion of $\mathcal{C}$. And $\hat{\mathcal{C}}$ denotes $\cap \mathcal{C}_\mu$, for all positive bounded measures μ.

The sets belonging to $\hat{\mathcal{C}}$ are called universally measurable. Remark that if μ is a σ-finite measure, then there exists a bounded measure which has the same negligible sets. Remark also that if μ is a σ-finite measure on $(T, \mathcal{C})$, then $\hat{\mathcal{C}}_\mu = \mathcal{C}_\mu$.

22 - The following theorem is due to Aumann [1] (in case of Lusin space. Extension to Suslin space is due to Sainte Beuve [24]).

Theorem III.22. Let $(T,\mathcal{T})$ be a measurable space and S a Suslin space. Let Γ be a multifunction from T to non empty subsets of S, whose graph G belongs to $\mathcal{T}\otimes\mathcal{B}(S)$. Then there exists a sequence (σ_n) of selections of Γ such that, for every t, $\{\sigma_n(t)\}$ is dense in $\Gamma(t)$, and σ_n is measurable for $\mathcal{T}$ and $\mathcal{B}(S)$. Moreover one can chose the σ_n such that : σ_n is the limit of a sequence $\mathcal{T}$ measurable functions assuming a finite number of values, and if in addition μ is a Radon measure on T (if T is a Hausdorff topological space) then σ_n is Lusin μ-measurable.

We shall prove theorems 22 and 23 simultaneously and in two stages (and we shall give another proof of theorem 23 in section 29 and another proof of theorem 22 when Γ is closed valued).

Application to functions (see also theorem 36).

Let $(T,\mathcal{T})$ be a measurable space, S a Suslin space and $u : T \to S$ a function. Consider the following properties :

(i) the graph of u belongs to $\mathcal{T}\otimes\mathcal{B}(S)$,

(ii) u is $(\mathcal{T}, \mathcal{B}(S))$ measurable,

(iii) u is the limit of a sequence of $\mathcal{T}$ measurable functions assuming a finite number of values,

(iv) (to be considered only if T is a Hausdorff topological space, and μ a Radon measure on T) u is Lusin μ-measurable.

Then $i \Rightarrow ii$, $i \Rightarrow iii$, $i \Rightarrow iv$ and $ii \Rightarrow i$.

If moreover S is completely regular, $iii \Rightarrow i$ and $iv \Rightarrow i$.

Proof. i $\Rightarrow$ ii, i $\Rightarrow$ iii and i $\Rightarrow$ iv follow from theorem 22.

Denote by $G(u)$ the graph of u, and by Δ_S the diagonal of S^2, by 1_S the identical map on S, and by $u \times 1_S$ the map $(t,x) \mapsto (u(t),x)$.

Then the fact that ii $\Rightarrow u \times 1_S$ is $(\mathcal{C} \otimes \mathcal{B}(S), \mathcal{B}(S) \otimes \mathcal{B}(S))$ measurable, and that $\Delta_S \in \mathcal{B}(S^2) = \mathcal{B}(S) \otimes \mathcal{B}(S)$, and the formula $G(u) = (u \times 1_S)^{-1} (\Delta_S)$ prove ii $\Rightarrow$ i.

Finally if S is completely regular, by lemma 31 below there exists a sequence (f_n) of real continuous functions on S which separate points of S. The formula

$$G(u) = \bigcap_n \{(t,x) \in T{\times}S \mid f_n(x) = f_n \circ u(t)\} \text{ holds}$$

But either of the hypotheses (iii) and (iv) implies that $f_n \circ u$ isis $\mathcal{C}$ measurable. Therefore iii $\Rightarrow$ i and iv $\Rightarrow$ i.

23 - The following theorem is called the projection theorem.

Theorem III.23. Let $(T,\mathcal{C})$ be a measurable space and S a Suslin space. If G belongs to $\mathcal{C} \otimes \mathcal{B}(S)$, its projection $pr_T(G)$ belongs to $\mathcal{C}$.

24 - Definition. A measurable space $(T, \mathcal{C})$ is said to be separable if there exists a sequence (A_n) in $\mathcal{C}$ which generates $\mathcal{C}$ and if the functions χ_{A_n} separate the points of T.

Lemma III.24. Let $(T,\mathcal{C})$ be a measurable space. If $\mathcal{C}$ is generated by the family $(A_i)_{i \in I}$, then for every t

$$\bigcap_i \{\theta \in T \mid \chi_{A_i}(\theta) = \chi_{A_i}(t)\} = \bigcap_{A \in \mathcal{C}} \{\theta \in T \mid \chi_A(\theta) = \chi_A(t)\} = \bigcap \{A \mid t \in A \text{ and } A \in \mathcal{C}\}.$$

Consequently, if $\mathcal{C}$ separates the points of T, so the family $(A_i)_{i \in I}$ separates the points of T.

Proof. The last equality is obvious. Let us prove the first one. Inclusion $\supset$ is obvious. Conversely if $\chi_{A_i}(\theta) = \chi_{A_i}(t)$ for every i, let us consider the set $\mathcal{B}$ of subsets B of T such that $\chi_B(\theta) = \chi_B(t)$. Then $\mathcal{B}$ is the σ-field $\mathcal{B} = \{B \in \mathcal{P}(T) | \{t,\theta\} \subset B$ or $\{t,\theta\} \subset T - B\}$, and contains every A_i. Hence $\mathcal{B}$ contains $\mathcal{C}$.

25 - Proposition III.25. If $(T,\mathcal{C})$ satisfies definition 24, the map $h : t \mapsto (\chi_{A_n}(t))_{n\in\mathbb{N}}$ from T to $\{0,1\}^{\mathbb{N}}$ is an isomorphism of measurable spaces between T and $h(T)$.

Proof. First h is one-to-one. As $t \mapsto \chi_{A_n}(t)$ is $\mathcal{C}$-measurable, so is h. It remains to prove that if $A \in \mathcal{C}$, $h(A)$ belongs to $\mathcal{B}(h(T))$. The set $\{A \in \mathcal{P}(T) | h(A) \in \mathcal{B}(h(T))\}$ is a σ-field. It contains A_n because

$$h(A_n) = \{(\xi_i) \in \{0,1\}^{\mathbb{N}} | \xi_n = 1\} \cap h(T).$$

26 - Lemma III.26. Let $(\Omega_1, \mathcal{A}_1)$ and $(\Omega_2, \mathcal{A}_2)$ be two measurable spaces, and $\varphi : \Omega_1 \to \Omega_2$ a measurable map. Then if μ is a positive bounded measure on Ω_1, φ is $(\mathcal{A}_{1\mu}, \mathcal{A}_{2\varphi(\mu)})$ measurable. Hence φ is $(\hat{\mathcal{A}}_1, \hat{\mathcal{A}}_2)$ measurable. Futhermore if Ω_2 is a Suslin space and $\mathcal{S}$ is the σ-field on Ω_2 generated by Suslin subsets, φ is $(\hat{\mathcal{A}}_1, \mathcal{S})$ measurable.

Proof. 1) Let $B \in \mathcal{A}_{2\varphi(\mu)}$. Then $B = B_0 \cup N$ with $B_0 \in \mathcal{A}_2$ and N is $\varphi(\mu)$ negligible. Then $\varphi^{-1}(B) = \varphi^{-1}(B_0) \cup \varphi^{-1}(N)$. But $\varphi^{-1}(N)$ is contained in a μ-negligible subset of $\mathcal{A}_1$ because N is contained in a $\varphi(\mu)$-negligible subset of $\mathcal{A}_2$. Thus $\varphi^{-1}(B) \in \mathcal{A}_{1\mu}$.

2) For every μ, $\varphi^{-1}(\hat{\mathcal{A}}_2) \subset \varphi^{-1}(\mathcal{A}_{2\varphi(\mu)}) \subset \mathcal{A}_{1\mu}$. Hence

$$\varphi^{-1}(\hat{\mathcal{A}}_2) \subset \cap \mathcal{A}_{1\mu} = \hat{\mathcal{A}}_1 .$$

3) If Ω_2 is a Suslin space, $\mathcal{S}$ is contained in $\hat{\mathfrak{a}}_2$, because every positive bounded measure ν on Ω_2 is Radon, and every Suslin subset is ν-measurable.

27 - Proof of theorems 22 and 23 when $(T,\mathcal{C})$ is separable.

By proposition 25 we may suppose T is a subspace of $E = \{0,1\}^{\mathbb{N}}$. Denote by $i: T \to E$ the embedding map.

It is easy to see (thanks to lemma 20) that

$$\begin{aligned} \mathcal{C} \otimes \mathcal{B}(S) &= (\mathcal{B}(E) \cap T) \otimes \mathcal{B}(S) \\ &= [\mathcal{B}(E) \otimes \mathcal{B}(S)] \cap [T \times S] . \end{aligned}$$

Therefore $G = G' \cap [T \times S]$ with $G' \in \mathcal{B}(E) \otimes \mathcal{B}(S)$.

As E is compact metrizable, G' is a Suslin subset of $E \times S$.

1) Let us prove theorem 23. We have

$$\mathrm{pr}_T\, G = \mathrm{pr}_E\,(G') \cap T = i^{-1}\,(\mathrm{pr}_E\, G').$$

The result follows from the last part of lemma 26, because $\mathrm{pr}_E\, G'$ is Suslin.

2) Let us prove now theorem 22. The set G' is the graph of a set valued function Γ' from $\mathrm{pr}_E\, G'$ to non empty subsets of S, such that $\Gamma'(t) = \Gamma(t)$ for every t. Theorem 18 applies to Γ' (with $T = \mathrm{pr}_E\, G'$, $X = S$, $\Gamma = \Gamma'$). Let σ'_n be the selections thus obtained. Denote by $j : T \to \mathrm{pr}_E\, G'$ the embedding map. Put $\sigma_n = \sigma'_n \circ j$. By lemma 26 j is $(\hat{\mathcal{C}}, \mathcal{S})$ measurable (we denote by $\mathcal{S}$ the σ-field generated by Suslin subsets of $\mathrm{pr}_E\, G'$). Consequently σ_n is $(\hat{\mathcal{C}}, \mathcal{B}(S))$ measurable and is the limit of a sequence of $\hat{\mathcal{C}}$-measurable functions assuming a finite number of values. We shall prove the Lusin measurability property in the next section.

28 - Proofs in general case.

First there exists a countably generated sub σ-field of $\mathcal{C}$, $\mathcal{C}_0$, such that $G \in \mathcal{C}_0 \otimes \mathcal{B}(S)$ (indeed, consider the set $\mathcal{E}$ of all countably generated sub σ-fields of $\mathcal{C}$. Then $\cup\{\mathcal{G} \otimes \mathcal{B}(S) | \mathcal{G} \in \mathcal{E}\}$ is a σ-field, hence it is equal to $\mathcal{C} \otimes \mathcal{B}(S)$).

Define an equivalence relation on T by $t \, R \, s$ is "for every $A \in \mathcal{C}_0$, $\chi_A(t) = \chi_A(s)$". Let $T_1 = T/R$ and $\varphi : T \to T_1$ be the canonical map. Every $A \in \mathcal{C}_0$ is a union of equivalence classes, thus if $A \neq B$, $\varphi(A) \neq \varphi(B)$ and if $A \cap B = \emptyset$, $\varphi(A) \cap \varphi(B) = \emptyset$. Therefore $\mathcal{C}_1 = \varphi(\mathcal{C}_0)$ is a tribe on T_1 (because $\emptyset \in \mathcal{C}_1$, the union of a sequence of elements of $\mathcal{C}_1$ belongs to $\mathcal{C}_1$, and $\varphi(T-A) = T_1 - \varphi(A)$ because $\varphi(T-A)$ and $\varphi(A)$ are disjoint and cover T_1). As $\mathcal{C}_0 \ni A \to \varphi(A) \in \mathcal{C}_1$ is injective and onto, $\mathcal{C}_1$ is countably generated. And $\mathcal{C}_1$ separates points of T_1 : indeed if $\varphi(t) \neq \varphi(s)$ there exists $A \in \mathcal{C}_0$ such that $t \in A$ and $s \notin A$, and then $\varphi(t) \in \varphi(A)$ and $\varphi(s) \in \varphi(T-A) = T_1 - \varphi(A)$. Hence, thanks to lemma 24, $(T_1, \mathcal{C}_1)$ is separable.

Consider now a set $H \in \mathcal{C}_0 \otimes \mathcal{B}(S)$. It is the graph of a multifunction Σ. We shall prove that $t \, R \, s$ implies $\Sigma(t) = \Sigma(s)$. First if $H = A \times B$, that is true because t and s belong or do not belong at the same time to A. Then the set of all $H \in \mathcal{P}(T \times S)$ having the property $t \, R \, s \Rightarrow \Sigma(t) = \Sigma(s)$ is a σ-field. Therefore it contains $\mathcal{C} \otimes \mathcal{B}(S)$.
We shall consider the multifunction Γ_1 from T_1 to subsets of S, defined by $\Gamma_1(\varphi(t)) = \Gamma(t)$. Let us prove that its graph G_1 belongs to $\mathcal{C}_1 \otimes \mathcal{B}(S)$. Any selection ψ of the multifunction φ^{-1} is $(\mathcal{C}_1, \mathcal{C}_0)$ measurable, because if $A \in \mathcal{C}_0$, $\psi^{-1}(A) = \varphi(A)$.

As $\Gamma_1(t_1) = \Gamma(\Psi(t_1))$ one has

$$\begin{aligned} G_1 &= \{(t_1,x) | x \in \Gamma_1(t_1)\} \\ &= \{(t_1,x) | (\Psi(t_1),x) \in G\} \\ &= (\Psi \times 1_S)^{-1} (G) \text{ belongs to } \mathcal{C}_1 \otimes \mathcal{B}(S). \end{aligned}$$

1) We prove now theorem 23. One has $pr_T(G) = \varphi^{-1}(pr_{T_1} G_1)$. By section 27 $pr_{T_1} G_1$ belongs to $\hat{\mathcal{C}}_1$. Hence $pr_T G \in \hat{\mathcal{C}}$ by lemma 26.

2) We prove now theorem 22. Section 27 applies to Γ_1. Let σ_n^1 be the selection thus obtained. Put $\sigma_n = \sigma_n^1 \circ \varphi$. Then σ_n is a selection of Γ, and for every t, $\{\sigma_n(t)\}$ is dense in $\Gamma(t)$. And thanks to lemma 26 σ_n is $(\hat{\mathcal{C}}, \mathcal{B}(S))$ measurable and is the limit of a sequence of $\hat{\mathcal{C}}$ measurable functions assuming a finite number of values.

It remains to prove that if T is a Hausdorff topological space and μ a Radon measure , the σ_n can be chosen Lusin μ-measurable. Let P be a Polish space and $h : P \to S$ continuous and onto. Denote by $1_T \times h$ the map $(t,p) \mapsto (t, h(p))$ from $T \times P$ to $T \times S$. Put $\Sigma(t) = h^{-1}(\Gamma(t))$. Then $\Sigma(t)$ is non empty, and the graph of Σ is $H = (1_T \times h)^{-1}(G)$ and belongs to $\mathcal{C} \otimes \mathcal{B}(P)$. Hence by previous results Σ has a sequence of measurable selections θ_n such that $\{\theta_n(t)\}$ is dense in $\Sigma(t)$. The θ_n (with values in P) are Lusin μ-measurable. Therefore $\sigma_n = h \circ \theta_n$ is also Lusin μ-measurable.

29 - <u>Remarks</u>. 1) We sketch another proof of theorem 23. When S is compact metrizable, this theorem is well known : G is analytic and its projection is also analytic and this implies $pr_T G \in \hat{\mathcal{C}}$ (see for example Meyer [18] for analytic sets). If S is Polish, S is G_δ (that is a countable intersection of open sets) in a compact metrizable space E. Then it is obvious that $\mathcal{C} \otimes \mathcal{B}(S) \subset \mathcal{C} \otimes \mathcal{B}(E)$ and theorem 23 is true for S polish.

Finally if S is Suslin let P be a Polish space and $h : P \to S$ continuous and onto. Then if $G \in \mathcal{C} \otimes \mathcal{B}(S)$, $(1_T \times h)^{-1}(G)$ belongs to $\mathcal{C} \otimes \mathcal{B}(P)$ and $pr_T\, G = pr_T\,[(1_T \times h)^{-1}(G)]$.

2) **If we add as hypothesis** of theorem 22 that Γ is closed valued, then theorem 22 is an easy consequence of theorems 23 and 8. Indeed let h be a continuous map from a Polish space P onto S, and put $\Sigma(t) = h^{-1}(\Gamma(t))$. Then the graph H of Σ is measurable, and if U is open in P $\Sigma^{-}(U) = pr_T(H \cap T \times U)$ belongs to $\hat{\mathcal{C}}$(th.23). So theorem 8 applies to Σ. Then the argument given at the end of section 28 remain valid.

30 - We can now extend theorem 9.

<u>Theorem III.30</u>. <u>Let</u> $(T, \mathcal{C}, \mu)$ <u>be a measurable space with</u> $\mu \geq 0$ <u>σ-finite</u> $\mathcal{C}$ <u>complete</u>, X <u>a complete separable metric space</u>, <u>and</u> Γ <u>map</u> T <u>to closed non empty subsets of</u> X. <u>Then the following properties are equivalent</u>

a) $\Gamma^{-}(U) \in \mathcal{C}$ <u>for every open set</u> U

b) $d(x, \Gamma(.))$ <u>is measurable for every</u> $x \in X$

c) Γ <u>admits a sequence of measurable selections</u> (σ_n) <u>such that</u> $\Gamma(t) = \overline{\{\sigma_n(t)\}}$

d) <u>the graph of</u> Γ, G, <u>belongs to</u> $\mathcal{C} \otimes \mathcal{B}(X)$,

e) $\Gamma^{-}(B) \in \mathcal{C}$ <u>for every borel set</u> B.

f) $\Gamma^{-}(F) \in \mathcal{C}$ <u>for every closed set</u> F.

<u>Proof</u>. By theorem 9 $a \Leftrightarrow b \Leftrightarrow c$.

By proposition 11 $f \Rightarrow a$, and by proposition 13 $b \Rightarrow d$. As $e \Rightarrow f$ is obvious, it remains to prove $d \Rightarrow e$. Remark that μ is equivalent to a bounded measure so applying definition 21 we obtain $\mathcal{C} \subset \hat{\mathcal{C}} \subset \mathcal{C}_\mu = \mathcal{C}$, Hence $\mathcal{C} = \hat{\mathcal{C}}$. But

$$\Gamma^{-}(B) = pr_T\,(G \cap [T \times B]) \text{ and } G \cap [T \times B] \text{ belongs to } \mathcal{C} \otimes \mathcal{B}(X).$$

So by theorem 23 $d \Rightarrow e$.

§5 - MEASURABILITY IN SUSLIN LOCALLY CONVEX SPACES

31 - The following lemma is due so Schwartz. The important **fact assumed** is that every family ofopen sets in E^2 has a countable subfamily which has the same union.

Lemma III.31. Let E be a Suslin topological space and $(f_i)_{i\in I}$ a family of real valued continuous functions which separates points of E (that is if $x \neq y$, there exists $i \in I$ such that $f_i(x) \neq f_i(y)$. Then there exists a countable subfamily $(f_i)_{i\in D}$ which still separates points of E.

Proof. The fact that $(f_i)_{i\in I}$ separates points of E is equivalent to

$$E^2 - \Delta_E = \bigcup_{i\in I} (f_i \times f_i)^{-1} (\mathbb{R}^2 - \Delta_{\mathbb{R}})$$

(in this formula $f_i \times f_i$ denotes the map $(x,y) \mapsto (f_i(x), f_i(y))$ and Δ_A the diagonal in $A\times A$). As E^2 is Suslin, there exists a Polish space P and a continuous onto map $h : P \to E^2$. **Let** $U_i = (f_i \times f_i)^{-1}(\mathbb{R}^2 - \Delta_{\mathbb{R}})$. It is an open set. It is well known that there exists a countable subset of I, D, such that

$$\bigcup_{i\in D} h^{-1}(U_i) = \bigcup_{i\in I} h^{-1}(U_i) .$$

As h is onto, that implies

$$\bigcup_{i\in D} U_i = \bigcup_{i\in I} U_i .$$

Hence the countable subfamily $(f_i)_{i\in D}$ separates points of E.

32 - Lemma III.32. Let E be a Hausdorff locally convex space and (e'_n) a sequence in E' which separates point of E (such a sequence exists if E is Suslin). Then the set of all linear combinations with rational coefficients of the e'_n is a countable dense subset of E' (endowed with $\tau(E',E)$).

<u>Proof</u>. Denote by H the set described in the statement. For any topology of topological vector space on E', the closure of H is a vector space. So if H_1 denotes the vector space generated by the e'_n, $\bar{H} = \bar{H}_1$. But as H_1 is convex its closure for $\tau(E',E)$ is the same as its closure for $\sigma(E',E)$. But for $\sigma(E',E)$, $\bar{H}_1 = E'$ because the orthogonal space $H_1^{\perp}$ is $\{0\}$.

33 - <u>Lemma III.33</u>. <u>Let</u> E <u>be a Hausdorff locally convex space and</u> E' <u>its dual endowed with the Mackey topology</u>. <u>Let</u> f <u>be a convex proper function on</u> E', <u>finite and continuous at one point</u> $x'_0 \in E'$. <u>Let</u> D <u>be a dense subset of</u> E'. <u>Then</u> $\inf \{f(y') | y' \in E'\} = \inf \{f(x') | x' \in D\}$.

<u>Proof</u>. The interior of dom f, $\overset{\circ}{\widehat{\text{dom } f}}$ is convex and non empty. So $D \cap \overset{\circ}{\widehat{\text{dom } f}}$ is dense in $\overset{\circ}{\widehat{\text{dom } f}}$. Moreover f is continuous on $\overset{\circ}{\widehat{\text{dom } f}}$. Thus for any $y' \in \overset{\circ}{\widehat{\text{dom } f}}$, $f(y') \geq \inf \{f(x') | x' \in D\}$.
Consider now $y' \in \text{dom } f - \overset{\circ}{\widehat{\text{dom } f}}$. It is well know (Bourbaki [3], II-2-6, prop. 16, p. 54) that $]y', x'_0[\subset \overset{\circ}{\widehat{\text{dom } f}}$. The restriction of f to $[y', x'_0]$ is finite and convex. Hence

$$\begin{aligned} f(y') &\geq \inf \{f(z') | z' \in]y', x'_0]\} \\ &\geq \inf \{f(z') | z' \in \overset{\circ}{\widehat{\text{dom } f}}\} \\ &\geq \inf \{f(z') | z' \in D\} . \end{aligned}$$

34 - <u>Lemma III.34</u>. <u>Let</u> E <u>be a Hausdorff locally convex space</u>, (x'_n) <u>a dense sequence in</u> E' <u>for</u> $\tau(E',E)$, <u>and</u> K <u>a closed convex weakly locally compact subset of</u> E <u>which contains no line</u>. <u>Then</u>

$$K = \bigcap_n \{x \in E | \langle x'_n, x\rangle \leq \delta^*(x'_n | K)\}.$$

Proof. Put $f = \delta^*(.|K)$. By I.15, lemma 33 applies to f with $D = \{x'_n | n \in \mathbb{N}\}$. We have

$$\begin{aligned}\delta(x|K) &= -\inf\{\delta^*(x'|K) - \langle x', x\rangle | x' \in E'\} \\ &= -\inf\{\delta^*(x'_n|K) - \langle x'_n, x\rangle | n \in \mathbb{N}\} \\ &= \sup\{\langle x'_n, x\rangle - \delta^*(x'_n|K) | n \in \mathbb{N}\}.\end{aligned}$$

So $x \in K \Leftrightarrow \delta(x|K) \le 0$

$$\Leftrightarrow \forall n,\ \langle x'_n, x\rangle - \delta^*(x'_n|K) \le 0 .$$

Remark. On that subject see Klee-Olech [14].

35 - Proposition III.35. *Let* E *be a Suslin locally convex space,* $(T, \mathcal{C}, \mu)$ *a measure space,* Γ *a multifunction from* T *to closed convex weakly locally compact subsets of* E *which contain no line,* Σ *from* T *to closed convex subsets of* E. *If* $\forall x' \in E'$, $\delta^*(x'|\Sigma(t)) \le \delta^*(x'|\Gamma(t))$ μ *almost everywhere, then* $\Sigma(t) \subset \Gamma(t)$ μ-*almost everywhere*.

Proof. By lemma 32 there exists a dense sequence (x'_n) in E' endowed with $\tau(E', E)$. For every n we have $\Sigma(t) \subset \{x \in E | \langle x'_n, x\rangle \le \delta^*(x'_n|\Gamma(t))\}$ μ-a.e.

So by lemma 34

$$\Sigma(t) \subset \Gamma(t) \quad \mu\text{-a.e.}$$

36 - Theorem III.36. *Let* $(T, \mathcal{C})$ *be a measurable space,* E *a Hausdorff locally convex space* S *a Suslin subset of* E *and* $\sigma : T \to S$. *Then the following properties are equivalent*

a) *for every* x', $\langle x', \sigma(.)\rangle$ *is* $\hat{\mathcal{C}}$-*measurable*.

b) σ *is the limit of a sequence of* $\hat{\mathcal{C}}$-*measurable functions assuming a finite number of values*.

c) *the graph of* σ *belongs to* $\hat{\mathcal{C}} \otimes \mathcal{B}(E)$

d) σ is $(\hat{\mathcal{T}}, \mathcal{B}(E))$ measurable

e) (to be considered only if T is a Hausdorff topological space and μ a Radon measure on T), σ is Lusin μ-measurable.

Proof. Remark first that c $\Leftrightarrow$ the graph of σ belongs to $\hat{\mathcal{T}} \otimes \mathcal{B}(S)$. The implications

c $\Rightarrow$ b, c $\Rightarrow$ d, c $\Rightarrow$ e follow from theorem 22.

And b $\Rightarrow$ a, d $\Rightarrow$ a, e $\Rightarrow$ a are obvious.

Finally a $\Rightarrow$ c follows from the formula

$$G = \bigcap_n \{(t,x) \in T \times S \mid \langle x'_n, x\rangle = \langle x'_n, \sigma(t)\rangle\}$$

where G is the graph of σ, and (x'_n) is a sequence in E' which separates points of S (lemma 31).

Remark. This theorem contains prop. 10, coroll. 2 and prop. 11 of Bourbaki [4] IV-5-5, p. 181-183.

37 - Theorem III.37. Let $(T, \mathcal{T})$ be a measurable space, E a Hausdorff locally convex space S a convex Suslin subset of E and Γ map T to closed convex non empty weakly locally compact subsets of S which contain no line. Then the following properties are equivalent :

a) for every x', $\delta^*(x' \mid \Gamma(.))$ is $\hat{\mathcal{T}}$-measurable.

b) Γ has a sequence of $\hat{\mathcal{T}}$ measurable (in sense of theorem 36) selections (σ_n) such that, for every t, $\Gamma(t) = \overline{\{\sigma_n(t) \mid n \in N\}}$

c) the graph of Γ belongs to $\hat{\mathcal{T}} \otimes \mathcal{B}(E)$.

Proof. First c $\Rightarrow$ b follows from theorem 22.

And b $\Rightarrow$ a follows from the formula

$$\delta^*(x' \mid \Gamma(t)) = \sup_n \langle x', \sigma_n(t)\rangle .$$

Note that the vector space F generated by S is still Suslin. Indeed $co(S \cup (-S)) = \varphi(S \times S \times [0,1])$ where $\varphi(x,y,\lambda) = \lambda x - (1-\lambda)y$.

So $co(S \cup (-S))$ is Suslin and $F = \bigcup_{n \in \mathbb{N}} n\, co(S \cup (-S))$ is also Suslin.

By lemmas 31 and 32 there exists a $\tau(F',F)$ dense sequence (y'_n) in F'.

Let G denotes the graph of Γ. As $\Gamma(t)$ is a closed convex weakly locally compact subset of F, it follows from lemma 34 that

$$G = \bigcap_n \{(t,x) \in T \times F \mid \langle y'_n, x \rangle \leq \delta^*(y'_n \mid \Gamma(t))\}$$

That proves $a \Rightarrow c$.

§6 - IMPLICIT FUNCTION THEOREM. STABILITY PROPERTIES OF MEASURABLE MULTIFUNCTIONS

38 - There are many theorems on existence of implicit measurable functions. We give only one theorem.

Theorem III.38. *Let* $(T, \mathcal{T})$, $(U, \mathcal{U})$ *be two measurable spaces,* S *a Suslin space,* Σ *a multifunction from* T *to non empty subsets of* S *whose graph belongs to* $\mathcal{T} \otimes \mathcal{B}(S)$, Θ *a multifunction from* T *to non empty subsets of* U *whose graph belongs to* $\mathcal{T} \otimes \mathcal{U}$ *and* $g : T \times S \to U$ *a* $(\mathcal{T} \otimes \mathcal{B}(S), \mathcal{U})$*-measurable map. If for every* t, $g(t, \Sigma(t)) \cap \Theta(t) \neq \emptyset$, *then there exists a* $(\hat{\mathcal{T}}, \mathcal{B}(S))$ *measurable selection of* Σ, σ, (*which is also the limit of a sequence of* $\hat{\mathcal{T}}$*-measurable functions assuming a finite number of values*) *such that* $g(t, \sigma(t)) \in \Theta(t)$.

Proof. Put

$$\Gamma(t) = \{x \in \Sigma(t) \mid g(t, x) \in \Theta(t)\}.$$

Then the graphs of Σ, Θ, Γ satisfy

$$G(\Gamma) = G(\Sigma) \cap \hat{g}^{-1}(G(\Theta))$$

where $\hat{g}$ denotes the map $(t, x) \mapsto (t, g(t, x))$ from $T \times S$ to $T \times U$.

As g is $(\mathcal{C}\otimes\mathcal{B}(S),\mathcal{C}\otimes\mathcal{U})$ measurable, $g^{-1}(G(\Theta))$ belongs to $\mathcal{C}\otimes\mathcal{B}(S)$. The existence of σ follows from theorem 22.

Application. In control theory $\Theta(t) = \{v(t)\}$ where $v : T \to U$ is a map ($v(t)$ is a velocity). If $v(t) \in g(t, \Sigma(t))$, then there exists a measurable selection of Σ, σ, such that $v(t) = g(t, \sigma(t))$. In the simplest case $\Sigma(t) = S$.

If S and U are metrizable and if $g(t, x)$ is continuous in x and measurable in t, then g is measurable. Indeed, as S is Suslin it is separable and lemma 14 applies.

39 - The following lemma is useful in the construction of new measurable multifunctions.

Lemma III.39. Let $(T,\mathcal{C})$ be a measurable space, S a Suslin space, $\varphi : T\times S \to \bar{\mathbb{R}}$ a $\mathcal{C}\otimes\mathcal{B}(S)$ measurable function and Σ a multifunction from T to subsets of S whose graph G belongs to $\mathcal{C}\otimes\mathcal{B}(S)$. Then

$$m(t) = \sup\ \{\varphi(t,x) \mid x \in \Sigma(t)\}$$

is a $\hat{\mathcal{C}}$-measurable function of t.

Proof. Let $\alpha \in \mathbb{R}$. Then

$$m(t) > \alpha \Leftrightarrow \exists\ x \in \Sigma(t) \ \text{ such that } \ \varphi(t,x) > \alpha .$$

Hence

$$\{m > \alpha\} = pr_T\ (G \cap \{(t,x) \mid \varphi(t,x) > \alpha\}).$$

By theorem 23 $\{m > \alpha\} \in \hat{\mathcal{C}}$.

Application. If for every t, the set

$$\Gamma(t) = \{x \in \Sigma(t) \mid \varphi(t,x) = m(t)\}$$

is non empty, Γ has a measurable graph (that is its graph belongs to $\hat{\mathcal{C}}\otimes\mathcal{B}(S)$) and has a measurable selection.

If the supremum is not obtained, there exist (except if $\Sigma(t)$ is empty) approaching values. For example if $m(t) \in \mathbb{R}$ and $\varepsilon > 0$, the set

$$\Gamma_\varepsilon(t) = \{x \in \Sigma(t) \mid \varphi(t,x) > m(t) - \varepsilon\} \qquad (\text{or} \geq)$$

is non empty and Γ_ε has a measurable selection.

40 - There are many theorems about stability of measurable multifunctions under various transformations. We shall give two theorems.

Theorem III.40. Let E be a complete separable metric space, $(\Gamma_n)_{n\in\mathbb{N}}$ a sequence of multifunctions from T to non empty subsets of E whose graphs belong to $\mathcal{C} \otimes \mathcal{B}(E)$. Then

1) the multifunctions $\cup \Gamma_n$ and $\cap \Gamma_n$ have measurable graphs. The multifunction $\bar{\Gamma}_0$ is $\hat{\mathcal{C}}$ measurable (in sense of theorem 30).

2) if moreover E is a separable Banach space, the multifunctions $\overline{\Gamma_1 + \Gamma_2}$ and $\overline{co}(\Gamma_1)$ are $\hat{\mathcal{C}}$-measurable (in sense of theorem 30).

Proof. 1) The first part of 1) is obvious. Consider now $\bar{\Gamma}_0$. By theorem 22 Γ_0 has a sequence of $\hat{\mathcal{C}}$ measurable selections (σ_n) such that $\{\sigma_n(t)\}$ is dense in $\Gamma_0(t)$. So by theorem 30, $\bar{\Gamma}_0$ is $\hat{\mathcal{C}}$ measurable.

2) Let $(\sigma_n^1)_{n\in\mathbb{N}}$ and $(\sigma_n^2)_{n\in\mathbb{N}}$ be sequences of $\hat{\mathcal{C}}$-measurable selections of Γ_1 and Γ_2 such that

$$\{\sigma_n^i(t) \mid n\in\mathbb{N}\} \text{ is dense in } \Gamma_i(t) .$$

Then $\overline{\Gamma_1(t) + \Gamma_2(t)} = \overline{\{\sigma_n^1(t) + \sigma_m^2(t) \mid (n,m) \in \mathbb{N}^2\}}$.

By theorem 30, $\overline{\Gamma_1 + \Gamma_2}$ is $\hat{\mathcal{C}}$-measurable.

Similarly the set of all barycenters with rational coefficients of finite subsets of $\{\sigma_n^1(t) \mid n\in\mathbb{N}\}$ is countable and dense in $\overline{co}(\Gamma_1(t)$.

- <u>Theorem III.41</u>. <u>Let</u> $(T, \mathscr{T})$ <u>be a measurable space</u>, <u>and</u> E <u>a separable Banach space</u>. <u>Let</u> $f : T \to E$ <u>be a measurable map and</u> $\rho : T \to [0, \infty[$ <u>a measurable function</u>. <u>Then</u>

1) $t \mapsto B(f(t), \rho(t))$ (<u>the closed ball with center</u> $f(t)$ <u>and radius</u> $\rho(t)$) <u>is a measurable multifunction</u> (<u>in sense of definition 10</u>).

2) <u>moreover if</u> Γ <u>from</u> T <u>to closed non empty subsets of</u> E <u>is measurable</u>, <u>then</u>

$$t \mapsto \Sigma(t) = \{x \in \Gamma(t) \mid \|f(t) - x\| = d(f(t), \Gamma(t))\}$$

<u>is</u> $\hat{\mathscr{T}}$<u>-measurable</u> (<u>and eventually empty valued</u>).

<u>Proof</u>. 1) Let (x_n) be a dense sequence in the unit ball of E. Put $\sigma_n(t) = f(t) + \rho(t)\, x_n$.

Then σ_n is measurable and

$$B(f(t), \rho(t)) = \overline{\{\sigma_n(t) \mid n \in N\}}.$$

By theorem 9, $t \mapsto B(f(t), \rho(t))$ is measurable.

2) The graphs of Σ and Γ verify

$$G(\Sigma) = G(\Gamma) \cap G(B(f(.), \rho(.))),$$

Where $\rho(t) = d(f(t), \Gamma(t))$.

So by theorem 30, Σ is $\hat{\mathscr{T}}$-measurable.

BIBLIOGRAPHY OF CHAPTER III

1 - AUMANN, R.J. - Measurable utility and measurable choice theorem. La décision C.N.R.S. (1967), p. 15-26.

2 - BOURBAKI, N. - Topologie générale ch. IX.

3 - BOURBAKI, N. - Espaces vectoriels topologiques ch. I - II.

4 - BOURBAKI, N. - Intégration ch. I - II - III - IV.

5 - BOURBAKI, N. - Intégration ch. IX.

6 - CASTAING, CH. - Sur les multi-applications mesurables. Revue Inf. Rech. Op. 1 (1967) - 91-126.

7 - CASTAING, CH. - Sur les multi-applications mesurables. Thèse Caen (1967).

8 - CASTAING, CH. - Intégrales convexes duales. C.R.A.S. 275 (1972), 1331-1334.

9 - DEBREU, G. - Integration of correspondences 5th Berkeley Symposium on Math. Stat. Prob. vol. II, part. I, p. 351-372.

10 - FILIPPOV, A.F. - On certain questions in the theory of optimal control. Siam J. Control, 1(1962), 76-84.

11 - HIMMELBERG, C.J. - JACOBS, M.Q. - VAN VLECK, F.S. - Measurable application, selectors and Filippov's implicit functions lemma. J. Math. An. Appl. 25-2 (1969), 276-284.

12 - HIMMELBERG, C.J. - VAN VLECK, F.S. - Some selection theorems for measurable functions. Can. J. Math. XXI - 2 (1969), 394-399.

13 - JACOBS, M.Q. - Measurable multivalued mappings and Lusin's theorem. Tr. A.M.S. 134-3 (1968), 471-481.

14 - KLEE, V. - OLECH, C. - Characterizations of a class of convex sets. Math. Scand. 20-2 (1967) 290-296.

15 - KURATOWSKI, K. - RYLL-NARDZEWSKI, C. - A general theorem on selectors Bull. Ac. Pol. Sc. 13 (1965), 397-403.

16 - LEESE, S.J. - Measurable selections in normed spaces. Proc. Edinburgh Math. Soc. 19(1974) 147-150.

17 - LEESE, S.J. - Multifunctions of Suslin type. Bull. Austr. Math. Soc. II (1974) 395-411.

18 - MEYER, P.A. - Probabilités et potentiel. Hermann (1966).

19 - NEVEU, J. - Bases mathématiques du calcul des probabilités. Masson (1964).

20 - PLIS, A. - Remark on measurable set-valued functions. Bull. Ac. Pol. Sc. 9-12 (1961), 857-859.

21 - ROCKAFELLAR, R.T. - Measurable dependance of convex sets and functions on parameters. J. Math. An. Appl. 28 (1969) 4-25.

22 - ROGERS, C.A. - WILLMOTT, R.C. - On the uniformization of sets in topological spaces. Acta Mathematica 120 - 1-2 (1968), 1-52.

23 - ROHLIN, V.A. - Selected topics from the metric theory of dynamical systems. Uspehi Mat. Nauk 4-2 (1949), 57-128 (in Russian). American Math. Soc. Translations. Vol. 49, série 2, p. 171-240 (in english).

24 - SAINTE BEUVE, M.F. - Sur la généralisation d'un théorème de section mesurable de von Neumann-Aumann. C.R.A.S. 276 (1973), 1297-1300, and : On the extension of von Neumann-Aumann's theorem. Journal of Funct. An. 17-1 (1974) 112-129.

25 - SION, M. - Uniformization of sets in topological spaces. Tr. A.M.S. 96 (1960) 237-245.

26 - VALADIER, M. - Contribution à l'Analyse Convexe. Thèse, Paris 1970.

27 - VALADIER, M. - Espérance conditionnelle d'un convexe fermé aléatoire. Séminaire d'Analyse Convexe, Montpellier 1972, exposé n° 1.

28 - VON NEUMANN, J. - On rings of operators. Reduction theory. Ann. of Math. 50 (1949), 401-485.

CHAPTER IV

TOPOLOGICAL PROPERTY OF THE PROFILE OF A MEASURABLE MULTIFUNCTION WITH COMPACT CONVEX VALUES

§1 - THE MAIN THEOREM AND ITS COROLLARIES

1 - The motivation to study the measurability of the profile (i.e. the set of extreme points) of a measurable multifunction with convex values comes from the theory of optimal control. The result we present here will enable us in fact to give a parametric version of Choquet's theorem and Caratheodory's theorem.

Finally, the parametric version of Caratheodory's theorem will be used to formulate the generalization of Ljapunov's theorem which was essentially our original motivation. Recently, Godet-Thobie uses the measurability of the profile in the theory of multimeasures ([11]).

2 -In this section, we denote by $(\Omega, \mathcal{G})$ an abstract measurable space, E a Hausdorff locally convex topological vector space and E' the dual vector space of E. The following result was communicated to the first author by J.J. Moreau and will be useful in the proof of our main theorem.

Proposition IV.2. *Let* A *be a non empty equicontinuous set of* E', f *a mapping from* A *to* $\mathbb{R}$; *suppose there exists a number* m *such that* $f(y) \leq m$ *for every* $y \in A$. *Then, the function* h *defined on* E *by*

$$x \mapsto h(x) = \sup \{f(y) - \langle y,x\rangle \mid y \in A\}$$

is finite and uniformly continuous on E.

Proof. It is clear that $h(x) > -\infty$ for all x in E. For x and v in E, we have

$$\begin{aligned} h(x-v) &= \sup \{f(y) - \langle v,y\rangle \mid y \in A\} \\ &= \sup \{f(y) - \langle y,x\rangle + \langle y,v\rangle \mid y \in A\} \\ &\leq \sup \{\langle y,v\rangle \mid y \in A\} + h(x) \end{aligned}$$

By taking $x=0$ in this inequality, we have

$$h(-v) \leq \sup \{\langle y,v\rangle \mid y \in A\} + m$$

Since A is equicontinuous, the function $v \mapsto \sup \{\langle y,v\rangle \mid y \in A\}$ is finite forevery v in E, and this function is uniformly continuous on E. Hence h is finite and uniformly continuous on E.

3 - *Theorem IV.3. Let* F *be a Hausdorff separable locally convex space, let* E *be the vector dual of* F *endowed with a locally convex topology which is compatible with the duality. Let* S *be a subspace of* E *and* $\mathcal{B}(S)$ *the Borel tribe of* S. *Let* f *be a mapping from* $\Omega\times S$ *to* $]-\infty,+\infty]$ *with the following conditions* :

1) *For every fixed* ω *in* Ω, *the domain* $\operatorname{dom} f_\omega$ *of* $f(\omega,.)$

$$\operatorname{dom} f_\omega = \{y \in S \mid f(\omega,y) < +\infty\}$$

is equicontinuous,

2) *For every fixed* ω *in* Ω, $f(\omega,.)$ *is continuous and majorized on* $\operatorname{dom} f_\omega$,

3) *There exists a sequence* (σ_n) *of scalarly* $\mathcal{G}$*-measurable mappings from* Ω *to* S *such that the sequence* $(\sigma_n(\omega))$ *is dense in* $\operatorname{dom} f_\omega$ *for every fixed* ω *in* Ω *and such that the functions* $\omega \mapsto f(\omega, \sigma_n(\omega))$ *are* $\mathcal{G}$*-measurable.*

Then the function

$$\hat{f} : (\omega,y) \mapsto \inf \{\langle x,y\rangle + \beta \mid \beta \in \mathbb{R},\ x \in F,\ \langle v,x\rangle + \beta \geq f(\omega,v),\ \forall\, v \in \operatorname{dom} f_\omega\}$$

is $\mathcal{G}\otimes\mathcal{B}(S)$*-measurable.*

<u>Proof</u>. For every ω in Ω and every y in S, we have

$$\hat{f}(\omega,y) = \inf \{\langle x,y\rangle + \beta_x(\omega) \mid x \in F\}$$

where $$\beta_x(\omega) = \sup \{f(\omega,v) - \langle v,x\rangle \mid v \in \operatorname{dom} f_\omega\}$$

for all x in F and all ω in Ω. It follows from condition 1) and proposition IV.2, that the function $(\omega,x) \mapsto \beta_x(\omega)$ is finite on $\Omega \times F$ and the mapping $x \mapsto \beta_x(\omega)$ is (uniformly) continuous on F for every fixed ω in Ω. Moreover, for every fixed x in F, the mapping $\omega \mapsto \beta_x(\omega)$ is $\mathcal{G}$-measurable since we have

$$\beta_x(\omega) = \sup_n \{f(\omega, \sigma_n(\omega)) - \langle \sigma_n(\omega), x\rangle\}$$

thanks to condition 2) and condition 3). Let (x_n) be a dense sequence in F. Then, it follows from the continuity of $x \mapsto \beta_x(\omega)$, that

$$\begin{aligned}\hat{f}(\omega,y) &= \inf \{\langle x,y\rangle - \beta_x(\omega) \mid x \in F\}\\ &= \inf_n \{\langle x_n,y\rangle - \beta_{x_n}(\omega)\}\end{aligned}$$

This proves that $\hat{f}$ is $\mathcal{G}-\mathcal{B}(S)$-measurable.

<u>Corollary IV.4. With hypotheses and notations of theorem IV.3., if the graph of the multifunction $\omega \mapsto \operatorname{dom} f_\omega$ belongs to $\mathcal{G} \otimes \mathcal{B}(S)$ and if the mapping f is $\mathcal{G} \otimes \mathcal{B}(S)$-measurable, then the graph of the multifunction</u>

$$\Phi : \omega \mapsto \{y \in \operatorname{dom} f_\omega \mid f(\omega,y) = \hat{f}(\omega,y)\}$$

<u>belongs to $\mathcal{G} \otimes \mathcal{B}(S)$.</u>

<u>Proof</u>. Since the graph of the multifunction $\omega \mapsto \operatorname{dom} f_\omega$ belongs to $\mathcal{G} \otimes \mathcal{B}(S)$ and since f and $\hat{f}$ are $\mathcal{G} \otimes \mathcal{B}(S)$-measurable, the graph of Φ belongs to $\mathcal{G} \otimes \mathcal{B}(S)$ too.

Corollary IV.5. With hypotheses of theorem IV.3, let Γ be a multifunction from Ω with non empty convex equicontinuous closed values in E. Assume there exists a sequence (σ_n) of scalarly $\mathcal{A}$-measurable mappings from Ω to E such that for every fixed ω in Ω the sequence $(\sigma_n(\omega))$ is dense in $\Gamma(\omega)$. Denote by $\Gamma(\omega)$ the set of extreme points of $\Gamma(\omega)$. Then the graph of the multifunction $\Gamma : \omega \to \Gamma(\omega)$ belongs to $\mathcal{A} \otimes \mathcal{B}(E)$.

Proof. Let (x_n) be a dense sequence in $F=E'$. Then the set $\mathcal{F}$ of all linear combinations of (x_n) is countable and dense in E' with respect to the Mackey topology $\tau(E',E)$ so that for every convex compact set K in E and for every point y in E with $y \notin K$, there exists an element e in $\mathcal{F}$ such that

$$\sup_{x \in K} <e,x> \ < \ <e,y>$$

Hence the graph $G(\Gamma)$ of Γ belongs to $\mathcal{A} \otimes \mathcal{B}(E)$ since we have

$$G(\Gamma) = \bigcap_{e \in \mathcal{F}} G(\Gamma_e)$$

where $\Gamma_e(\omega) = \{y \in E \mid <e,y> \leq \sup_{u \in \Gamma(\omega)} <e,u>\}$

As the functions

$$\omega \mapsto \sup_{u \in \Gamma(\omega)} <e,u> = \sup_n <e, \sigma_n(\omega)>$$

are $\mathcal{A}$-measurable for all e in E', we deduce easily that the sets

$$G(\Gamma_e) = \{(\omega,y) \in \Omega \times E \mid y \in \Gamma_e(\omega)\}$$

belong to $\mathcal{A} \otimes \mathcal{B}(E)$, so that $G(\Gamma)$ belongs to $\mathcal{A} \otimes \mathcal{B}(E)$ too. As $\mathcal{F}$ is countable, we can put $\mathcal{F} = \{e_m\}_{m \in \mathbb{N}}$. It is clear that the functions

$$\omega \mapsto h_m(\omega) = \sup \{ |\langle e_m, y\rangle| \; y \in \Gamma(\omega)\}$$
$$= \sup_n \{ |\langle e_m, \sigma_n(\omega)\rangle| \}$$

are $\mathfrak{G}$-measurable for all $m \in \mathbb{N}$. Let us consider the following mapping $f : \Omega \times E \to [0,+\infty]$

$$f(\omega,s) = \begin{cases} \sum_{m=1}^{\infty} \dfrac{\langle e_m, s\rangle^2}{m^2(1+h_m(\omega))^2} & \text{if } s \in \Gamma(\omega) \\ +\infty \quad \text{if } s \notin \Gamma(\omega) \end{cases}$$

It is clear that f satisfies the conditions of theorem IV.2 (by taking $S=E$, $\operatorname{dom} f_\omega = \Gamma(\omega)$). In addition $f(\omega,.)$ is continuous and strictly convex on $\Gamma(\omega)$ because the sequence (e_m) separates the points of E. The corollary IV.4 then shows that the graph of the multifunction

$$\Phi : \omega \mapsto \{y \in \Gamma(\omega) | \; f(\omega,y) = \hat{f}(\omega,y)\}$$

belongs to $\mathfrak{G} \otimes \mathcal{B}(E)$. By virtue of a well known characterization of the set of extreme points of a convex compact set (see for instance [1], p. 32), $\Phi(\omega)$ is equal to the set $\Gamma(\omega)$ of extreme points of $\Gamma(\omega)$. This finishes the proof.

Remark. a) The existence of the mapping f given in the corollary IV.5 is a parametric version of Hervé's theorem ([1] , theor. I.4.3.,p.32) which shows the existence of a strictly convex continuous function defined on a convex compact metrizable set of a Hausdorff locally convex space.

b) The result given in Corollary IV.5 was stated by Benamara ([3.2]) by different methods.

The arguments we utilized in the preceding paragraphs allow us to obtain the following variants of the results given in these paragraphs.

Theorem IV.6. *Let T be a metric space and let Γ be a continuous multifunction from T with non empty convex compact values in a Hausdorff locally convex space E. Assume there exists in the dual vector E' of E a sequence (e'_n) which separate the points of E. Denote by $\ddot{\Gamma}(t)$ the set of extreme points of $\Gamma(t)$. Then the graph $G(\dot{\Gamma})$ of the multifunction $t \mapsto \ddot{\Gamma}(t)$ $(t \in T)$ is a G_δ subset of the graph of Γ.*

Proof. As Γ is continuous, it is not hard to check that, for every continuous function φ defined on E, the function

$$t \mapsto \sup \{\varphi(x) \mid x \in \Gamma(t)\} \quad (t \in T).$$

is continuous (see for instance Berge ([4])).

Let us put

$$h_n(t) = \sup \{|\langle e'_n, x\rangle| \mid x \in \Gamma(t)\} \quad (t \in T)$$

$$f(t,x) = \begin{cases} \sum_{n=1}^{\infty} \dfrac{\langle e'_n, x\rangle^2}{n^2(1+h_n(t))^2} & \text{if } x \in \Gamma(t) \\ +\infty \text{ if } x \notin \Gamma(t) \end{cases}$$

It follows from the continuity of the multifunction Γ that the graph of Γ is closed in $T\times E$. Hence, f is lower-semi-continuous on $T\times E$. In addition, the function $f(t,.)$ is continuous and strictly convex on $\Gamma(t)$. Let us prove now that the function

$$\hat{f}(t,x) = \inf \{\langle x,y\rangle + \beta \mid \beta \in \mathbb{R},\ y' \in E',\ \langle v,y'\rangle + \beta \geq f(t,v),\ \forall v \in \Gamma(t)\}$$

is upper-semi-continuous on $T \times E$. Indeed, we have

$$\hat{f}(t,x) = \inf \{\langle x,y'\rangle + \beta_{y'}(t) \mid y' \in E'\}$$

where

$$\beta_{y'} = \sup \{f(t,v) - \langle y',v\rangle \mid v \in \Gamma(t)\}$$

It suffices to show that $\beta_{y'}$ is upper-semi-continuous on T. First we remark that, for every compact set K in T, the graph of the restriction $\Gamma_{|K}$ of Γ to K is compact since $\Gamma(K)$ is compact (see Berge ([4]). So, it is enough to prove that the restriction of $\beta_{y'}$ to every compact set K in T is upper-semi-continuous. If $G(\Gamma_{|K})$ is the graph of $\Gamma_{|K}$, then f is continuous on $G(\Gamma_{|K})$. Let $r \in \mathbb{R}$. We have to verify that the set

$$\{t \in K \mid \beta_{y'}(t) \geq r\}$$

is compact. Let (t_k) be a sequence in K such that $\beta_{y'}(t_k) \geq r$ for all k and such that the sequence (t_k) converges to $t \in K$. There exists a sequence (v_k) such that

$$\begin{cases} v_k \in \Gamma(t_k) \\ f(t,v_k) - \langle y',v_k\rangle \geq r \,. \end{cases}$$

Since the sequence (v_k) is relatively compact, we can suppose that the sequence (v_k) converge to v in $\Gamma(t)$ since $G(\Gamma_{|K})$ is compact. This implies that

$$\lim_{k \to \infty} f(t,v_k) - \langle y',v_k\rangle = f(t,v) - \langle y',v\rangle \geq r$$

Hence $\beta_{y'}(t) \geq r$. We deduce easily that $\hat{f}$ is upper-semi-continuous. As $\hat{f}-f$ is upper-semi-continuous on $T \times E$ it follows that

$$G(\ddot{\Gamma}) = \bigcap_n \{(t,x) \in G(\Gamma) \mid \hat{f}(t,x) - f(t,x) < \tfrac{1}{n}\}$$

is a G_δ subset of $G(\Gamma)$. This achieves the proof.

7 - We give now a variant of theorem IV.6.

Theorem IV.7. Let T be a topological space and let X be a convex compact set of a Hausdorff locally convex space E. Let Γ be an upper-semi-continuous multifunction from T with non empty convex compact values in X. If f is a real continuous function defined on X, then the function

$$(t,x) \mapsto \hat{f}(t,x) = \inf \{\langle x,y\rangle + \beta \mid \beta \in \mathbb{R},\ y \in E',\ \langle y,v\rangle + \beta \geq f(v), \forall v \in \Gamma(t)\}$$

is upper-semi-continuous on $T \times X$.

Proof. Denote by $\mathcal{M}_+^1(X)$ the set of positive Radon measure on X with mass 1. It is well known that $\mathcal{M}_+^1(X)$ is convex and compact with respect to the vague topology. Put

$$\Sigma(t) = \{\mu \in \mathcal{M}_+^1(X) \mid \mu(\Gamma(t)) = 1\},\ \forall\ t \in T$$

Then the support function of the convex compact set $\Sigma(t)$ of $\mathcal{M}_+^1(X)$ is given by

$$\delta^*(h, \Sigma(t)) = \sup \{h(x) \mid x \in \Gamma(t)\}$$

where h belongs to the space $\mathcal{C}(X)$ of real continuous functions on X. Moreover by a well known result ([1], corollary I.3.6., p. 26).

$$\hat{f}(t,x) = \sup \{\mu(f) \mid \mu \in \Sigma(t),\ b_\mu = x\}$$

where b_μ is the barycenter of measure μ. Since the mapping $\mu \mapsto b_\mu$ from $\mathcal{M}_+^1(X)$ to X is clearly continuous, we conclude that the multifunction

$$\psi : x \mapsto \{\mu \in \mathcal{M}_+^1(X) \mid b_\mu = x\} \quad (x \in X)$$

is upper-semi-continuous from X with convex compact values in $\mathcal{M}_+^1(X)$. Now, by a result of Berge ([4]), the function $t \mapsto \delta^*(h, \Sigma(t))$ is upper-semi-continuous on T for every fixed h in $\mathcal{C}(X)$. Hence the multifunction Σ is upper-semi-continuous from T with convex compact values in $\mathcal{M}_+^1(X)$.

Finally, the multifunction

$$(t,x) \mapsto \Sigma(t) \cap \psi(x)$$

from $T \times X$ with convex compact values in $\mathcal{K}^1_+(X)$ is upper-semi-continuous. Since $\mu \mapsto \mu(f)$ is continuous on $\mathcal{M}^1_+(X)$, we apply again Berge's result to obtain the upper-semi-continuity of

$$(t,x) \mapsto \hat{f}(t,x) = \sup \{\mu(f) \mid \mu \in \Sigma(t) \cap \psi(x)\}$$

8 - Corollary IV.8. With the hypotheses of theorem IV.7, let $G(\Gamma)$ be the graph of Γ. Then, the following set

$$\{(t,x) \in G(\Gamma) \mid \hat{f}(t,x) = f(x)\}$$

is a G_δ subset of $G(\Gamma)$

Proof. This follows directly from the upper-semi-continuity of $\hat{f}$ because we have

$$\{(t,x) \in G(\Gamma) \mid \hat{f}(t,x) = f(x)\} = \bigcap_{n \in N} \{(t,x) \in G(\Gamma) \mid \hat{f}(t,x) - f(t) < \frac{1}{n}\}$$

9 - Corollary IV.9. Under the hypotheses of corollary IV.8, assume that T and X are metrizable. Then the graph $G(\ddot{\Gamma})$ of $\ddot{\Gamma}$ is a G_δ subset of $T \times X$.

Proof. If X is metrizable, there exists a strictly convex continuous func f on X ([1], theorem. I.4.3., p. 32) such that

$$\ddot{\Gamma}(t) = \{x \in \Gamma(t) \mid \hat{f}(t,x) = f(x)\}$$

Hence, by corollary IV.8 and the metrizability of $T \times X$, the graph $G(\ddot{\Gamma})$ of $\ddot{\Gamma}$,

$$G(\ddot{\Gamma}) = \{(t,x) \in G(\Gamma) \mid \hat{f}(t,x) = f(x)\}$$

is a G_δ subset of $T \times X$.

Remark. In corollary IV.9, we have given with the help of theorem IV.7 another proof of the topological property of the profile of a multifunction.

§2 - APPLICATIONS

Parametric version of Caratheodory's theorem.

For the convenience of the reader, we give first a lemma of measurability we use in the proof of the parametric version of Caratheodory's theorem.

10 - Lemma IV.10. *Let $(\Omega, \mathfrak{G})$ and $(X, \mathfrak{X})$ be two abstract measurable spaces. Let Γ_i $(i=1,2)$ be a multifunction from Ω with no empty values in X such that its graph $G(\Gamma_i)$ belongs to $\mathfrak{G} \otimes \mathfrak{X}$. Then, the graph $G(\Gamma)$ of the multifunction*

$$\omega \mapsto \Gamma(\omega) = \Gamma_1(\omega) \times \Gamma_2(\omega) \quad (\omega \in \Omega)$$

from Ω with no empty values in the product space $X \times X$ belongs to $\mathfrak{G} \otimes \mathfrak{X} \otimes \mathfrak{X}$.

Proof. Let Δ be the diagonal of $\Omega \times \Omega$ and let

$C = \{(\omega, x, \omega, y) \in \Omega \times X \times \Omega \times X \mid (\omega, x, y) \in \Omega \times X \times X\}$.

Let φ be the one to one mapping from C to $\Omega \times X \times X$ which is deduded from the canonical mapping from Δ onto Ω. Then we have

$$G(\Gamma) = \varphi[C \cap G(\Gamma_1) \times G(\Gamma_2)]$$

Since $G(\Gamma_1) \times G(\Gamma_2)$ belongs to $\mathfrak{G} \otimes \mathfrak{X} \otimes \mathfrak{G} \otimes \mathfrak{X}$, $C \cap G(\Gamma_1) \times G(\Gamma_2)$ belongs to the σ-algebra $\mathfrak{C}$ induced by the σ-algebra $\mathfrak{G} \otimes \mathfrak{X} \otimes \mathfrak{G} \otimes \mathfrak{X}$ on the set C. Let us show that φ is a measurable isomorphism from the measurable space $(C, \mathfrak{C})$ onto the measurable space $(\Omega \times X \times X, \mathfrak{G} \otimes \mathfrak{X} \otimes \mathfrak{X})$. Indeed, if A belongs to $\mathfrak{G}$, and B_1 and B_2 belong to $\mathfrak{X}$, then

$$\varphi^{-1}(A \times B_1 \times B_2) = (A \times B_1 \times A \times B_2) \cap C$$

this shows that φ is measurable. Now, given (A_i) $(i=1,2)$ in $\mathcal{G}$ and (B_i) $(i=1,2)$ in $\mathcal{X}$, we have

$$\varphi[(A_1 \times B_1 \times A_2 \times B_2) \cap C] = (A_1 \cap A_2) \times (B_1 \times B_2)$$

Whence, φ^{-1} is measurable. Consequently, $G(\Gamma)$ belongs to $\mathcal{G} \otimes \mathcal{X} \otimes \mathcal{X}$.

Remark. This lemma is due to Godet-Thobie ([11]) and is an analogue version of one of Castaing's results ([7.1], lemma 2.4) which shows that the graph of the product $\Gamma(\cdot) = \prod_n \Gamma_n(\cdot)$ of a sequence of multifunctions Γ_n such that $G(\Gamma_n)$ is Suslin, is Suslin too.

Theorem IV.11. *Let $(\Omega, \mathcal{G}, \mu)$ be a σ-finite positive and complete measure space, Γ a $\mathcal{G}$-measurable multifunction from Ω with non empty convex compact values in $\mathbb{R}^n$ and u a $\mathcal{G}$-measurable mapping from Ω into $\mathbb{R}^n$ such that $u(\omega) \in \Gamma(\omega)$ for all $\omega \in \Omega$. Then there exists a $\mathcal{G}$-measurable mapping λ from Ω into the simplex Λ_{n+1}*

$$\Lambda_{n+1} = \{(\lambda^i) \mid \lambda^i \geq 0, \sum_{i+1}^{n+1} \lambda^i = 1\}$$

and $n+1$ $\mathcal{G}$-measurable mappings $u_1, u_2, \ldots, u_{n+1}$ from Ω to $\mathbb{R}^n$ such that

$$\begin{cases} u_i(\omega) \in \ddot{\Gamma}(\omega), \ i=1,2,\ldots,n+1 \\ u(\omega) = \sum_{i+1}^{n+1} \lambda^i(\omega)\, u_i(\omega) \end{cases}$$

for all $\omega \in \Omega$.

Proof. Consider the multifunction

$$\Phi(\omega) = \Lambda_{n+1} \times (\ddot{\Gamma}(\omega))^{n+1} \qquad (\omega \in \Omega)$$

Let $\mathcal{B}(\mathbb{R}^n)$ be the Borel tribe of $\mathbb{R}^n$. By corollary IV.5, the graph $G(\ddot{\Gamma})$ of the multifunction $\ddot{\Gamma}$ belongs to $\mathcal{G}\otimes\mathcal{B}(\mathbb{R}^n)$. Hence, by virtue of lemma IV.10, the graph of the multifunction ϕ belongs to $\mathcal{G}\otimes\mathcal{B}(\Lambda_{n+1}\times(\mathbb{R}^n)^{n+1})$ where $\mathcal{B}(\Lambda_{n+1}\times(\mathbb{R}^n)^{n+1})$ is the Borel tribe of $\Lambda_{n+1}\times(\mathbb{R}^n)^{n+1}$. Define the continuous mapping from $\Lambda_{n+1}\times(\mathbb{R}^n)^{n+1}$ to $\mathbb{R}^n$ by

$$h(\lambda^1,\ldots,\lambda^{n+1}\ ;\ x_1,\ldots,x_{n+1}) = \sum_{i=1}^{n+1} \lambda^i x_i$$

Then by Caratheodory's theorem, we have

$$\Gamma(\omega) = h(\phi(\omega))\ ,\quad \forall\ \omega\in\Omega\ .$$

Therefore we can apply the measurable implicit function theorem III.38 to obtain the conclusion.

12 - Parametric version of Choquet's theorem.

Before stating the parametric version of Choquet's theorem, we need the following result of measurability due to Sainte Beuve([23], theorem 6, p. 125). We refer the reader to ([6.2], Bourbaki) for the theory of Radon measures on topological spaces.

Theorem IV.12. Let $(\Omega,\mathcal{G})$ be an abstract measurable space, $(X,\mathcal{B}(X))$ be a completely regular Suslin space with its Borel tribe $\mathcal{B}(X)$. Let Γ be a multifunction from Ω with non empty values in X such that its graph $G(\Gamma)$ belongs to $\mathcal{G}\otimes\mathcal{B}(X)$. Denote by $\mathcal{M}^1_+(X)$ the set of probability Radon measures on X equipped with the narrow topology and let $\mathcal{B}(\mathcal{M}^1_+(X))$ be the Borel tribe of the space $\mathcal{M}^1_+(X)$. Then the graph of the multifunction Σ,

$$\Sigma(\omega) = \{\lambda\in\mathcal{M}^1_+(X)\ |\ \lambda(\Gamma(\omega)) = 1\}\quad (\omega\in\Omega)$$

belongs to $\mathcal{G}\otimes\mathcal{B}(\mathcal{M}^1_+(X))$.

Proof. For every fixed $\omega \in \Omega$, denote by ε_ω the probability measure concentrated at the point ω and for every fixed G in $\mathcal{A} \otimes \mathcal{B}(X)$ define the mapping ψ_G from $\Omega \times \mathcal{M}_+^1(X)$ to $\mathbb{R}$ by

$$\psi_G(\omega, \lambda) = (\varepsilon_\omega \otimes \lambda)(G), \quad (\omega, \lambda) \in \Omega \times \mathcal{M}_+^1(X)$$

Let $\mathcal{H}$ be the elements F in $\mathcal{A} \otimes \mathcal{B}(X)$ such that the mappings ψ_F are $\mathcal{A} \otimes \mathcal{B}(\mathcal{M}_+^1(X))$-measurable. Given A in $\mathcal{A}$ and B in $\mathcal{B}(X)$ we have

$$\psi_{A \times B}(\omega, \lambda) = \chi_A(\omega) \times \lambda(B) = \chi_A(\omega) \times \varphi_B(\lambda)$$

where χ_A is the characteristic function of A and φ_B is defined by

$$\varphi_B(\lambda) = \lambda(B) \quad , \quad \lambda \in \mathcal{M}_+^1(X)$$

We assert first that the mapping φ_B is Borel. Indeed, **denote by** $\mathcal{B}_1$ the elements C of $\mathcal{B}(X)$ such that φ_C is Borel. Then the open sets of X belong to $\mathcal{B}_1$, because for every open set U in X, the characteristic function χ_U is lower-semi-continuous and bounded, so that the function $\lambda \mapsto \varphi_C(\lambda) = \langle \chi_U, \lambda \rangle$ is lower-semi-continuous on $\mathcal{M}_+^1(X)$ ([6.2], prop. 6, p. 59). Now, if U is an open set and V is **a closed set** in X, then the function $\lambda \mapsto \varphi_{U \cap V}(\lambda)$ is Borel too since we have

$$\varphi_{U \cap V} = \varphi_U - \varphi_{U \cap V^c}$$

where V^c **is the complement of** V; this proves that for every finite sequence of open (resp. closed) set U_i (resp. V_i) ($i = 1, 2, \ldots, n$), $\bigcup_{i+1}^{n}(U_i \cap V_i)$ belongs to B_1 and B_1 is a monotone class, so that $B_1 = \mathcal{B}(X)$ (see Neveu [21], théorème 19, p. 27). Hence φ_B is Borel for all $B \in \mathcal{B}_1(X)$. It follows that the mapping

$$\psi_{A \times B} : (\omega, \lambda) \mapsto \chi_A(\omega) \times \varphi_B(\lambda)$$

is $\mathcal{G}\otimes\mathcal{B}_1(X)$-measurable for all A in $\mathcal{G}$ and for all B in $\mathcal{B}_1(X)$.

Then for every finite sequence of elements $(A_j \times B_j)$ $(j=1,\dots,p)$ in $\mathcal{G}\otimes\mathcal{B}_1(X)$, $\bigcup_{j=1}^{p} A_j \times B_j$ belongs to $\mathcal{H}$ which is a monotone class. By ([21], theoreme 19, p. 27) we conclude that $\mathcal{H} = \mathcal{G}\otimes\mathcal{B}(X)$. Let

$$G_1 = \{(\omega,\lambda) \in \Omega \times \mathcal{M}_+^1(X) \mid (\epsilon_\omega \otimes \lambda)(G(\Gamma)) = 1\}$$
$$= \psi_{G(\Gamma)}^{-1}(\{1\}) = \{(\omega,\lambda) \in \Omega + \mathcal{M}_+^1(X) \mid \lambda(\Gamma(\omega)) = 1\}$$

Then G_1 belongs to $\mathcal{G}\otimes\mathcal{B}(\mathcal{M}_+^1(X))$.

<u>Remarks</u> 1) If $\Gamma(\omega)$ is compact for every ω in Ω, one can prove the measurability of the multifunction Σ by a different way which yields **at the same** time the lower-semi-continuity (resp. upper-semi-continuity) of Σ if Γ is lower-semi-continuous (resp. upper-semi-continuous) on a topological space. The proofs of these facts are very simple in these cases. First, assume that $(\Omega,\mathcal{G},\mu)$ is a complete and σ-finite positive measure and Γ is a multifunction from Ω with non empty compact values in X such that the graph $G(\Gamma)$ of Γ belongs to $\mathcal{G}\otimes\mathcal{B}(X)$. Recall that the set $\mathcal{M}_+^1(X)$ of bounded positive Radon measures defined on X is a Suslin space if we endow $\mathcal{M}^1(X)$ with the narrow topology, i.e. the topology of pointwise convergence on the space $\mathcal{C}^b(X)$ of all real bounded <u>continuous functions defined on</u> X ([6.2]) ; in addition, for every compact subspace K of the space X, the narrow topology and the vague topology coincide on the set

$$\{\nu \in \mathcal{M}_+^1(X) \mid \nu(K) = 1\}$$

Now, it is clear that, for every $\omega \in \Omega$,

$$\Sigma(\omega) = \{\nu \in \mathcal{M}_+^1(X) \mid \nu(\Gamma(\omega)) = 1\}$$

is convex and compact with respect to the (narrow = vague) topology. So we can evaluate the support functional of the convex compact set $\Sigma(\omega)$:

$$\forall\, f \in \mathcal{C}^b(X),\ \delta^*(f, \Sigma(\omega)) = \sup\,\{f(x) \mid x \in \Gamma(\omega)\}$$

Let (σ_n) be a countable $\mathcal{G}$-measurable dense selection of Γ (cf. theorem em III.22). Then,

$$\delta^*(f, \Sigma(\omega)) = \sup_n f(\sigma_n(\omega))$$

This proves that $\omega \mapsto \delta^*(f, \Sigma(\omega))$ is $\mathcal{G}$-measurable for all f in $\mathcal{C}^b(X)$. Since $\mathcal{M}^1_+(X)$ is convex and Suslin with respect to the narrow topology, we conclude by lemma III.31 that

$$G(\Sigma) = \bigcap_{p=1}^{\infty} \{(\omega,\nu) \in \Omega \times \mathcal{M}^1_+(X) \mid \langle\nu, f_p\rangle = \delta^*(f_p, \Sigma(\omega))\}$$

where (f_p) is a sequence in $\mathcal{C}^b(X)$ which separates the points of $\mathcal{M}^1(X)$. So $G(\Sigma)$ belongs to $\mathcal{G} \otimes \mathcal{B}(\mathcal{M}^1_+(X))$. Second, we pass to the case where Γ is lower-semi-continuous (resp. upper-semi-continuous) from a topological space Z with non empty compact values in X. Denote by $\mathcal{M}(X)$ the space of all Radon measures on X equipped with the narrow topology $\sigma(\mathcal{M}(X), \mathcal{C}^b(X))$.

Let us prove that the multifunction

$$z \mapsto \Sigma(z) = \{\nu \in \mathcal{M}^1_+(X) \mid \nu(\Gamma(z)) = 1\} \quad (z \in Z)$$

is lower-semi-continuous (resp. upper-semi-continuous) on Z. Since $\Sigma(z)$ is convex compact for the narrow topology $\sigma(\mathcal{M}(X), \mathcal{C}^b(X))$ as we have observed above, it suffices (thanks to the theorems II.21 and II.20) to prove that the function

$$z \mapsto \delta^*(f,\Sigma(z)) = \sup \{\langle\nu,f\rangle \mid \nu \in \Sigma(z)\} = \sup \{f(u) \mid u \in \Gamma(z)\}$$

is lower-semi-continuous (resp. upper-semi-continuous) on Z, for every $f \in \mathcal{C}^b(X)$. Since it is well known that $z \mapsto \sup \{f(x) \mid x \in \Gamma(z)\}$ is lower-semi-continuous (resp. upper-semi-continuous)if Γ is lower-semi-continuous (resp. upper-semi-continuous) by Berge's result ([4]), this achieves the proof of our assertion.

2) The motivation to study the properties of the multifunction Σ comes from the theory of optimal controls ([12], [26]) and the theory of differential games. In chapter V, we utilize the property of Σ to state some compactness theorems which were used to obtain a result of desintegration of measures ([7.4]).

13 - Theorem IV.13. Let $(\Omega, \mathcal{G}, \mu)$ be a σ-finite positive measure space with $\mathcal{G}$ μ-complete, let X be a convex compact metrizable set of a Hausdorff locally convex space E and let Γ be a measurable multifunction of Ω with nonempty convex compact values in X. If u is a measurable mapping of Ω into X such that $u(\omega) \in \Gamma(\omega)$ for every $\omega \in \Omega$, then there exists a measurable mapping λ of Ω into the set $\mathcal{M}^1_+(X)$ of positive Radon measure on X with total mass 1 ($\mathcal{M}^1_+(X)$ is equipped with the vague topology for which $\mathcal{M}^1_+(X)$ is a compact metrizable space) such that, for every $\omega \in \Omega$, we have

$$\begin{cases} \lambda_\omega(\ddot{\Gamma}(\omega)) = 1 \\ b_{\lambda_\omega} = u(\omega) \end{cases}$$

where b_{λ_ω} is the barycenter of the measure λ_ω.

Proof. We give here the proof given by Sainte Beuve ([23], prop. 4, p. 126) ; however there is another proof due to Valadier which furnishes a stronger result ([25]. Consider the multifunction

$$\Phi(\omega) = (\mu \in \mathcal{M}^1_+(X) \mid \mu(\ddot{\Gamma}(\omega)) = 1,\ b_\mu = u(\omega)\}$$

Then, $\Phi(\omega)$ is equal to the intersection of the following sets

$$\Delta(\omega) = \{\mu \in \mathcal{M}^1_+(X) \mid \mu(\ddot{\Gamma}(\omega)) = 1\}$$

$$\Sigma(\omega) = \{\mu \in \mathcal{M}^1_+(X) \mid b_\mu = u(\omega)\}$$

where b_μ is the barycenter of μ. Moreover, by using corollary IV.7 we know that $G(\ddot{\Gamma})$ belongs to $\mathcal{G} \otimes \mathcal{B}(X)$, where $\mathcal{B}(X)$ is the Borel tribe of X ; then by virtue of theorem IV.13, the graph $G(\Delta)$ of the multifunction Δ belongs to $\mathcal{G} \otimes \mathcal{B}(\mathcal{M}^1_+(X))$ where $\mathcal{B}(\mathcal{M}^1_+(X))$ is the Borel tribe of the compact metrizable space $\mathcal{M}^1_+(X)$. Moreover, the mapping $\mu \mapsto b_\mu$ is clearly continuous from $\mathcal{M}^1_+(X)$ to X ; so the graph $G(\Sigma)$ of the multifunction Σ belongs to $\mathcal{G} \otimes \mathcal{B}(\mathcal{M}^1_+(X))$. Therefore the graph $G(\Phi)$ of Φ belongs also to $\mathcal{G} \otimes \mathcal{B}(\mathcal{M}^1_+(X))$. By Choquet's theorem ([20], p. 182, theorem 25), $\Phi(\omega)$ is not empty for every $\omega \in \Omega$. Then Φ admits by theorem III.22 a measurable selection λ with the required properties.

§3 - CHARACTERIZATION OF THE PROFILE OF A CONVEX SET OF MEASURABLE SELECTIONS

In this section, $(\Omega, \mathcal{G}, \mu)$ is a positive σ-finite measure space with $\mathcal{G}$ -μcomplete and E is a Hausdorff locally convex space. Let S be a convex Suslin subset of E and $\mathcal{B}(S)$ the Borel tribe of S. Let Γ be a multifunction from Ω with no empty convex values in E such that $\Gamma(\omega) \subset S$ for every $\omega \in \Omega$ and its graph $G(\Gamma)$ belongs to $\mathcal{G} \otimes \mathcal{B}(S)$. Let $\ddot{\Gamma}(\omega)$ be the profile of $\Gamma(\omega)$ and denote by $\mathcal{S}_\Gamma$ (resp. $\mathcal{S}_{\ddot{\Gamma}}$) the set of $\mathcal{G}$ -measurable mappings σ from Ω to S (i.e. $\sigma^{-1}(B) \in \mathcal{G}$ for every $B \in \mathcal{B}(S)$) such that $\sigma(\omega) \in \Gamma(\omega)$ (resp. $\sigma(\omega) \in \ddot{\Gamma}(\omega)$) a.e. Let S_Γ (resp. $S_{\ddot{\Gamma}}$) be the quotient of $\mathcal{S}_\Gamma$ (resp. $\mathcal{S}_{\ddot{\Gamma}}$) for the equivalence "equality almost everywhere".

We first give the following theorem which provides the characterization of the profile of the convex set S_Γ.

- Theorem IV.14. **Under the proceeding hypotheses,** let u be a $\mathcal{A}$-measurable selection of Γ. Then there exists a $\mathcal{A}$-measurable mapping $\tilde{u}$ from Ω to the Suslin space $Z = \frac{1}{2}(S-S)$ (i.e., $\tilde{u}^{-1}(B)$ belongs to $\mathcal{A}$ for every Borel set B in Z) such that

   (i) $u(\omega) \pm \tilde{u}(\omega) \in \Gamma(\omega)$ for all $\omega \in \Omega$

   (ii) $\tilde{u}(\omega) \neq 0$ if $u(\omega) \in \Gamma(\omega) \setminus \overset{..}{\Gamma}(\omega)$

Proof. We first prove that the set

$$A = \{\omega \in \Omega \mid u(\omega) \in \Gamma(\omega) \setminus \overset{..}{\Gamma}(\omega)\}$$

belongs to $\mathcal{A}$. Consider the continuous mapping φ from $S \times S$ to S,

$$\varphi(x,y) = \frac{x+y}{2}, \; x \in X, \; y \in S$$

and let Δ be the diagonal of $S \times S$. Put

$$\Phi(\omega) = \{(x,y) \in \Gamma(\omega) \times \Gamma(\omega) \setminus \Delta \mid \varphi(x,y) = u(\omega)\}$$

Then, we have

$$A = \{\omega \in \Omega \mid \Phi(\omega) \neq \emptyset\}$$

By lemma IV.10, the graph of the multifunction $\omega \mapsto \Gamma(\omega) \times \Gamma(\omega)$ belongs to $\mathcal{A} \otimes \mathcal{B}(S) \otimes \mathcal{B}(S) = \mathcal{A} \otimes \mathcal{B}(S^2)$. So the graph of the multifunction Φ belongs to $\mathcal{A} \otimes \mathcal{B}(S^2)$. Therefore, by projection theorem III.23, A belongs to $\mathcal{A}$. Now, we can apply selection theorem III.22 to obtain a $\mathcal{A}$-measurable mapping v from A to S such that $v(\omega) \in \Phi(\omega)$ for all $\omega \in \Omega$. Hence, **there exist two** $\mathcal{A}$-measurable mappings u_1 and u_2 from A to S such that, for every $\omega \in \Omega$,

$$\begin{cases} u_1(\omega) \neq u_2(\omega) \\ u_i(\omega) \in \Gamma(\omega), \; i=1,2 \\ u_1(\omega) + u_2(\omega) = 2\, u(\omega) \end{cases}$$

Let us define

$$\tilde{v}_i(\omega) = \begin{cases} u_i(\omega), & \omega \in A \\ u(\omega), & \omega \in \Omega \setminus A \end{cases} \qquad (i = 1,2)$$

$$\tilde{u} = \frac{1}{2}(\tilde{v}_1 - \tilde{v}_2)$$

Then $\tilde{v}_i(\omega) \in \Gamma(\omega)$ for all $\omega \in \Omega$ $(i = 1,2)$ and $\tilde{u}(\omega) \neq 0$ for all $\omega \in A$.

Since we have

$$u = \frac{1}{2}(\tilde{v}_1 + \tilde{v}_2)$$

we deduce easily that $u(\omega) \pm \tilde{u}(\omega) \in \Gamma(\omega)$ for all ω.

This finishes the proof.

15 - The construction given in the proof of theorem IV.14. is utilized to obtain the characterization of the profile of S_Γ. Namely we have the following theorem.

Theorem IV.15. Under the hypotheses of theorem IV.14, the profile of S_Γ is equal to $S_\Gamma^{\cdot\cdot}$.

Proof. For notation convenience, we will not distinguish between a mapping from Ω to S and its equivalence class under "equality almost everywhere".

Clearly we have $S_\Gamma^{\cdot\cdot} \subset \ddot{S}_\Gamma$. To prove the inverse inclusion, suppose that $u \in \ddot{S}_\Gamma \setminus S_\Gamma^{\cdot\cdot}$. With the construction given in the proof of theorem IV.14, the set

$$A = \{\omega \in \Omega \mid u(\omega) \notin \ddot{\Gamma}(\omega)\}$$

is $\mathfrak{G}$-measurable with $\mu(A) > 0$. In addition, by the same construction, there exist two measurable selections u_1 and u_2 of the restriction to A of the multifunction Γ such that

$$\begin{cases} u_1(\omega) \neq u_2(\omega) \\ 2\,u(\omega) = u_1(\omega) + u_2(\omega) \end{cases}$$

for all $\omega \in A$. Put

$$\tilde{v}_i(\omega) = \begin{cases} u_i(\omega) & \text{if } \omega \in \Omega \\ u(\omega) & \text{if } \omega \in \Omega \setminus A \end{cases} \qquad (i=1,2)$$

Then $\tilde{v}_i \in S_\Gamma$ and we have

$$\begin{cases} \tilde{v}_1 + \tilde{v}_2 = 2u \\ \tilde{v}_1 \neq \tilde{v}_2 \end{cases}$$

This proves that $u \notin \ddot{S}_\Gamma$ contradicting the extreme nature of u.

16 - The following theorem describes a stronger aspect of theorem IV.14 and generalizes a theorem due to Ito ([15]).

Theorem IV.16. Let T be a compact metrizable space and $\mathcal{B}(T)$ its Borel tribe. Let E be a Hausdorff locally convex space. X a convex compact metrizable subset of E. Let Γ be an upper-semi-continuous multifunction from T with non empty convex compact values in X and $\ddot{\Gamma}(t)$ the set of extreme points of $\Gamma(t)$. Then we have the following properties.

1) The graph $G(\Gamma_0)$ of the multifunction

$$\Gamma_0(t) = \Gamma(t) \setminus \ddot{\Gamma}(t) \qquad (t \in T)$$

is a Borel set of $T \times X$.

2) There exists a Borel mapping φ from $G(\Gamma_0)$ to the compact metrizable space $Z = \frac{1}{2}(X-X)$ such that

$$\begin{cases} x \pm \varphi(t,x) \in \Gamma(t) \\ \varphi(t,x) \neq 0 \,. \end{cases}$$

for all $(t,x) \in G(\Gamma_0)$.

Proof. 1) By corollary IV.9, we know that the graph of the multifunction $t \mapsto \ddot{\Gamma}(t)$ is a G_δ subset of $T \times X$, so that

$$G(\Gamma_0) = G(\Gamma) \cap \complement G(\ddot{\Gamma})$$

is a countable union of compact subsets of $T\times X$, hence $G(\Gamma_0)$ is a Borel set of $T\times X$.

2) For every $t \in T$ and $x \in X$, define

$$\Sigma(t,x) = \{z \in \frac{1}{2}(X-X) \mid (x+z,\ x-z) \in \Gamma(t) \times \Gamma(t)\}$$

Clearly the graph of the multifunction Σ is closed in $T\times X\times Z$. Consider the multifunction Σ_0 defined on $G(\Gamma_0)$ by

$$\Sigma_0(t,x) = \Sigma(t,x) \setminus \{0\}$$

Then $\Sigma(t,x)$ is not empty for all $(t,x) \in G(\Gamma_0)$. But $\complement_Z\{0\}$ is a countable union of compact sets say, $\complement_Z\{0\} = \bigcup_n K_n$, so we have

$$\Sigma_0(t,x) = \bigcup_n (\Sigma(t,x) \cap K_n),\ (t,x) \in G(\Gamma_0)$$

Since $\{(t,x) \in G(\Gamma_0) \mid \Sigma(t,x) \cap K_n \neq \emptyset\}$ is a Borel set, then, by a standard procedure, we construct a Borel partition of $G(\Gamma_0)$, say, (B_p) such that $\Sigma(t,x) \cap K_p \neq \emptyset$ for every $(t,x) \in B_p$. For each p, the multifunction $(t,x) \mapsto \Sigma(t,x) \cap K_p$ defined on B_p admits a Borel selection σ_p by theorem III.6. Clearly, the mapping $\varphi : G(\Gamma_0) \to Z$

$$\varphi(t,x) = \sigma_p(t,x) \quad \text{for} \quad (t,x) \in B_p$$

satisfies the required properties.

Remark. An inspection of the proof of theorem IV.16 shows that this theorem still holds if Γ is a $\mathcal{G}$-measurable multifunction from an abstract measurable space $(\Omega, \mathcal{G})$ with no empty convex compact values in a separable Fréchet space. The details of proof are left to the reader.

§4 - EXTENSION OF LJAPUNOV'S THEOREM

17 - Definition. Let $(\Omega,\mathcal{G})$ be an abstract measurable space and let $m = (\mu_1,\dots,\mu_p)$ be a vector measure defined on $\mathcal{G}$ with values in $\mathbb{R}^n$ such that the measure space $(\Omega,\mathcal{G},|m|)$ is σ-finite with $\mathcal{G}$ $|m|$-complete. The measure m is semi-convex if for every $|m|$-integrable mapping v from Ω to $\mathbb{R}^n$, and for every $\mathcal{G}$-measurable set A belonging to $\mathcal{G}$, there exists a $\mathcal{G}$-measurable $A_0 \subset A$ such that

$$\int_{A_0} v \otimes dm = \frac{1}{2}\int_A v \otimes dm$$

Remark. If $|m|$ is non atomic, it is well known that m is semi-convex ([13], [19]). Namely, the vector measure n

$$n(B) = \int_B v \otimes dm, \quad B \in \mathcal{G}$$

satisfies the following convexity condition (see Appendix). For every fixed $A \in \mathcal{G}$, the set $\{n(B) \mid B \subset A,\ B \in \mathcal{G}\}$ is convex in $\mathbb{R}^{np}$. Then, given $\alpha \in [0,1]$ and A, remark that the empty set ϕ is $\mathcal{G}$-measurable and $\alpha\, n(A) = \alpha\, n(A) + (1-\alpha)\, n(\phi) \in \{n(B) \mid B \subset A,\ B \in \mathcal{G}\}$. So, there exists $A_0 \subset A$ with $A_0 \in \mathcal{G}$ such that $\alpha\, n(A) = n(A_0)$.

Theorem IV.17. Under the preceding hypotheses and notations, let Γ be a $\mathcal{G}$-measurable multifunction from Ω with no empty convex compact values in $\mathbb{R}^n$. Suppose there exists a positive $|m|$-integrable function h such that $\Gamma(\omega) \subset h(\omega) B$ for every $\omega \in \Omega$, where B is the unit ball of $\mathbb{R}^n$. Let us denote by $\mathcal{S}_\Gamma$ (resp. $\mathcal{S}_{\ddot{\Gamma}}$) the set of $|m|$-integrable mappings f from Ω to $\mathbb{R}^n$ such that $f(\omega) \in \Gamma(\omega)$ (resp. $f(\omega) \in \ddot{\Gamma}(\omega)$) a.e. . If m is semi-convex, then

$$\{\int f \otimes dm \mid f \in \mathcal{S}_\Gamma\} = \{\int f \otimes dm \mid f \in \mathcal{S}_{\ddot{\Gamma}}\}$$

and these sets are convex and compact in $\mathbb{R}^{np}$.

Proof. First we note that the quotient S_Γ of $\mathcal{S}_\Gamma$ by the equivalence "equality almost everywhere" is convex and not empty by virtue of theorem III.6 and the hypothesis. Moreover S_Γ is compact(*) for the topology $\sigma(L^1_{\mathbb{R}^n}(\Omega,\mathcal{G},|m|),\ L^\infty_{\mathbb{R}^n}(\Omega,\mathcal{G},\ |m|))$. Since the mapping $T : f \mapsto \int f \otimes dm$ from $L^1_{\mathbb{R}^n}(\Omega,\mathcal{G},|m|)$ to $\mathbb{R}^{np}$ is linear and continuous, $T(S_\Gamma)$ is convex compact. Now, given f_0 in S_Γ let

$$T^{-1}(f_0) = \{f \in S_\Gamma \mid \int f \otimes dm = \int f_0 \otimes dm\}$$

Then, $T^{-1}(f_0)$ is a convex $\sigma(L^1_{\mathbb{R}^n}(\Omega,\mathcal{G},|m|),\ L^\infty_{\mathbb{R}^n}(\Omega,\mathcal{G},|m|))$ compact set of S_Γ. To prove our theorem, it suffices to show that the extreme points of $T^{-1}(f_0)$ belong to $S^{\cdot\cdot}_\Gamma$. Let u_0 be an extremal point of $T^{-1}(f_0)$ and suppose $u_0 \notin S^{\cdot\cdot}_\Gamma$. By theorem IV.14, there exists a $\mathcal{G}$-measurable set A with $|m|(A) > 0$ and a $\mathcal{G}$-measurable mapping v from A to $\mathbb{R}^n$ such that

$$\begin{cases} u_0(\omega) \pm v(\omega) \in \Gamma(\omega) & , \quad \forall\, \omega \in A \\ v(\omega) \neq 0 & , \quad \forall\, \omega \in A \end{cases}$$

Since m is semi-convex, there exists a $\mathcal{G}$-measurable set $A_0 \subset A$ such that

$$\int_{A_0} v \otimes dm = \frac{1}{2} \int_A v \otimes dm$$

Define

$$\tilde{u}(\omega) = \begin{cases} v(\omega) & \omega \in A \\ -v(\omega) & \omega \in A \setminus A_0 \\ 0 & \omega \notin A \end{cases}$$

(*) The compactness of S_Γ is easy to prove here by using for instance Dunford-Pettis theorem's (see Dunford-Schwartz [9], p. 298). A simple proof of this property is established in the more general case in chapter V.

Then

$$u_o(\omega) = \frac{u_o(\omega) + \tilde{u}(\omega)}{2} + \frac{u_o(\omega) - \tilde{u}(\omega)}{2}, \quad \forall\, \omega \in \Omega$$

$$u_o(\omega) \pm \tilde{u}(\omega) \in \Gamma(\omega), \quad \forall\, \omega \in \Omega$$

By a simple calculation we have

$$\int \tilde{u} \otimes dm = 0$$

so that $u_o \pm \tilde{u} \in T^{-1}(f_o)$. Since u_o is a convex combination of $u_o + \tilde{u}$ and $u_o - \tilde{u}$, this contradicts the extremal property of u_o.

Remarks a) The idea of the proof is already given by Karlin([14]) and Karlin-Studden ([15]). The crucial fact is the construction of ũ such that

$$\begin{cases} u_o(\omega) + \tilde{u}(\omega) \in \Gamma(\omega), & \forall\, \omega \in \Omega \\ u_o(\omega) - \tilde{u}(\omega) \in \Gamma(\omega), & \forall\, \omega \in \Omega \\ \int_A \tilde{u} \otimes dm = 0 \quad \text{where } A = \{\omega \in \Omega \mid u(\omega) \in \Gamma(\omega) \setminus \ddot{\Gamma}(\omega)\} \end{cases}$$

b) We used in our proof the semi-convexity of the measure m. If $|m|$ is non atomic, then m is semi-convex by Ljapunovs's convexity result: this is, however, the easier result ([12], [18] whose proof is given in the appendix of this chapter and hence our proof of theorem IV.17 is a new proof of the extension of Ljapunov's theorem ([5], [7.3], [14], [15], [22], [25]). Now, it is interesting to give another proof of theorem IV.17 based on Caratheodory's parametric theorem IV.11.

Second proof of theorem IV.17

The idea of the proof is simple. First, we prove theorem IV.17 in the particular case where $\Gamma(\omega) = \Lambda_n = \{(\lambda^i) \mid \lambda^i \geq 0, \sum_{i=1}^{n} \lambda^i = 1\}$ for all $\omega \in \Omega$ and $|m|$ is bounded. Second, we apply Caratheodory's parametric theorem IV.11 to prove the general case.

First step : $\Gamma(\omega) = \Lambda_n$, $\forall\ \omega \in \Omega$; $|m|\ (\Omega) < +\infty$.

Since Λ_n is a convex compact set of $\mathbb{R}^n$, it is easily verified that S_Γ is $\sigma\,(L^\infty_{\mathbb{R}^n}\,(\Omega, \mathcal{G}, |m|), L^1_{\mathbb{R}^n}\,(\Omega, \mathcal{G}, |m|))$ compact and convex. Indeed, S_Γ is clearly convex and $\sigma(L^2_{\mathbb{R}^n}(\Omega, \mathcal{G}, |m|), L^2_{\mathbb{R}^n}(\Omega, \mathcal{G}, |m|))$, compact hence S_Γ is $\sigma(L^\infty_{\mathbb{R}^n}\,(\Omega, \mathcal{G}, |m|), L^1_{\mathbb{R}^n}(\Omega, \mathcal{G}, |m|))$ compact because m is bounded. Let $f_o \in S_\Gamma$ and put

$$T^{-1}(f_o) = \{f \in S_\Gamma \mid \int f \otimes dm = \int f_o \otimes dm\}$$

Then, $T^{-1}(f_o)$ is convex and $\sigma(L^\infty_{\mathbb{R}^n}, L^1_{\mathbb{R}^n})$ compact because the mapping $T : f \mapsto \int f \otimes dm$ is linear and weakly continuous from $L^\infty_{\mathbb{R}^n}$ to $\mathbb{R}^{np}$.

Now we proceed exactly in the same manner as in the proof of theorem IV.17, bu- we do not use the selection theorems. Let u_o be an extremal point of $T^{-1}(f_o)$. We have to prove that u_o belongs to $S^{\cdot\cdot}_\Gamma$. It suffices to check that the $\mathcal{G}$-measurable set A for which $0 < u^i_o(\omega) < 1$ for at least one i $(i = 1,\dots,n)$ is a null set, i.e., $|m|\ (A) = 0$. Suppose $|m|\ (A) > 0$. Then there exist two integers i_o and i_1, a number $\epsilon \in\ ]0,1[$ and a $\mathcal{G}$ -measurable set B such that

$$\begin{cases} \epsilon < u^i(\omega) < 1-\epsilon & \text{for } x \in B \text{ and } i=i_o,i_1 \\ |m|\ (B) > 0 \end{cases}$$

Since m is semi-convex, there exists a $\mathcal{G}$-measurable set $B_o \subset B$ such that

$$m(B_o) = \tfrac{1}{2}\, m(B)$$

Define

$$\tilde{u}(\omega) = \begin{cases} \tilde{u}^{i_o} = -\ \tilde{u}^{i_1} = -\ \epsilon & \text{for } \omega \in B_o \\ \tilde{u}^{i_o} = -\ \tilde{u}^{i_1} = \epsilon & \text{for } \omega \in B \setminus B_o \\ \tilde{u}^{i_o} = \tilde{u}^{i_1} = 0 & \text{for } \omega \in \Omega \setminus B \\ \tilde{u}^{i} = 0\ , \ i \neq i_o, i_1\ , & \text{for } \omega \in \Omega \end{cases}$$

Then, a simple calculation yields

$$u_o \pm \tilde{u} \in S_\Gamma$$
$$\int \tilde{u} \otimes dm = 0$$

So, $u_o \pm \tilde{u} \in T^{-1}(f_o)$. But, clearly

$$u_o = \frac{u_o + \tilde{u}}{2} + \frac{u_o - \tilde{u}}{2}$$

This contradicts the extremal nature of u_o.

<u>Second step : The general case</u>

Let $u_o \in S_\Gamma$. By theorem IV.11, we have

$$u_o(\omega) = \sum_{i=1}^{n+1} \lambda^i(\omega)\ u^i(\omega) , \quad \forall\ \omega \in \Omega$$

where the mapping $\omega \mapsto (\lambda^i(\omega))_{1 \le i \le n+1}$ is $\mathcal{G}$-measurable with $\lambda^i(\omega) \ge 0$ and $\sum_{i=1}^{n+1} \lambda^i(\omega) = 1$ for all ω and $u_i \in S^{..}_\Gamma$ $(i=1,\ldots,n+1)$. By integrating, we obtain

$$\int u_o \otimes dm = \sum_{i=1}^{n+1} \int (\lambda^i\ u_i) \otimes dm$$

By applying the result of the first step to the constant multifunction $\Lambda_{n+1} = \{(\lambda^i) \mid \lambda^i \ge 0,\ \sum_{i=1}^{n+1} \lambda^i = 1\}$ and the bounded vector measure $(u_i\ \mu_j)_{\substack{1 \le i \le n+1 \\ 1 \le j \le p}}$ there exists $\tilde{\lambda} \in S^{..}_{\Lambda_{n+1}}$ such that

$$\int \lambda^i\ u_i\ d\mu_j = \int \tilde{\lambda}^i\ u_i\ d\mu_j ,\quad i=1,\ldots,n+1\ ;\ j=1,\ldots,p .$$

Put $A_i = \{\omega \in \Omega \mid \tilde{\lambda}^i(\omega) = 1\}$, $(i=1,\ldots,n+1)$. Then the sets A_i are $\mathcal{G}$-measurable and disjoint and their union is Ω. Define

$$v(\omega) = u_i(\omega) \quad \text{for} \quad \omega \in A_i .$$

Then $v \in S^{..}_\Gamma$ and clearly

$$\int u_o \otimes dm = \sum_{i=1}^{n+1} \int (\tilde{\lambda}^i\ u_i) \otimes dm = \int v \otimes dm$$

That completes the proof.

Remark. There are many different proofs of theorem IV.17 (see for instance Olech and Valadier). However, the construction given in the first proof shows at the same time that $S^{\cdot\cdot}_\Gamma$ is dense in S_Γ for the weak topology $\sigma(L^1_{\mathbb{R}^n}, L^\infty_{\mathbb{R}^n})$ (This property will be established in Chapter V).

18 - Now, we give in the following the parametric version of theorem IV.17

Theorem IV.18. Let $m = (\mu_1, \ldots, \mu_p)$ *be a Radon vector measure on a locally compact Polish space* Ω. *Let* Γ *be a* $|m|$*-measurable multifunction from* Ω *with non empty convex compact values in* $\mathbb{R}^n$. *Suppose* $\Gamma(\omega) \subset h(\omega)\,B$ *for* $\omega \in \Omega$, *where* h *is a positive* $|m|$*-integrable function and* B *the unit ball of* $\mathbb{R}^n$. *Denote by* S_Γ *(resp.* $S^{\cdot\cdot}_\Gamma$*) the set of all* $|m|$*-integrable mappings* f *from* Ω *to* $\mathbb{R}^n$ *with* $f(\omega) \in \Gamma(\omega)$ *(resp.* $f(\omega) \in \Gamma^{\cdot\cdot}(\omega)$*) a.e. . Then*

1) $S^{\cdot\cdot}_\Gamma$ *is a* G_δ *subset of the convex compact set* S_Γ *endowed with the topology* $\sigma(L^1_{\mathbb{R}^n}(\Omega, \mathcal{G}, |m|), L^\infty_{\mathbb{R}^n}(\Omega, \mathcal{G}, |m|))$.

2) *Let* (T, ν) *be a locally compact* σ*-finite measure space where* ν *is a positive Radon measure on* T *and let* g *be an element of* $L^\infty_{\mathbb{R}^n}(\Omega \times T, |m| \otimes \nu)$. *Suppose that* m *is semi-convex. Let* φ *be a* ν*-measurable mapping from* T *to the Polish space* $(S_\Gamma, \mathcal{B}(S_\Gamma))$ *where* $\mathcal{B}(S_\Gamma)$ *is the Borel tribe of* S_Γ. *Then there exists a* ν*-measurable mapping* $\tilde{\varphi}$ *from* T *to the Polish space* $(S^{\cdot\cdot}_\Gamma, \mathcal{B}(S^{\cdot\cdot}_\Gamma))$ *where* $\mathcal{B}(S^{\cdot\cdot}_\Gamma)$ *is the Borel tribe of* $S^{\cdot\cdot}_\Gamma$ *such that*

$$\int \langle \tilde{\varphi}_t, g_t \rangle \, dm = \int \langle \varphi_t, g_t \rangle \, dm \quad \text{for all} \quad t \in T$$

Proof. 1) Since $S^{\cdot\cdot}_{\Gamma}$ is the set of extreme points of S_{Γ} (theorem IV.15), $S^{\cdot\cdot}_{\Gamma}$ is clearly a G_{δ} subset of the convex compact metrizable space S_{Γ} because $L^1_{\mathbb{R}^n}(\Omega,|m|)$ is separable.

2) By hypothesis, for every u in $L^{\infty}_{\mathbb{R}^n}(\Omega,|m|)$, the mapping $t \mapsto \langle\varphi_t, u\rangle$ is ν-measurable and for every fixed t in T, the mapping $u \mapsto \langle\varphi_t, u\rangle$ is continuous on $L^{\infty}_{\mathbb{R}^n}(\Omega, |m|)$ with respect to the weak topology $\sigma(L^{\infty}_{\mathbb{R}^n}(\Omega,|m|), L^1_{\mathbb{R}^n}(\Omega,|m|))$. Remark that the mapping $t \to g_t$ from T to $L^{\infty}_{\mathbb{R}^n}(\Omega,|m|)$ is clearly ν-measurable with respect to this topology because $L^1_{\mathbb{R}^n}(\Omega,|m|)$ is a separable Banach space and the dual space of $L^1_{\mathbb{R}^n}(\Omega,|m|)$ is $L^{\infty}_{\mathbb{R}^n}(\Omega,|m|)$ and, since the weak dual of a separable Banach space is a Lusin space, then, by applying the preceding remarks and theorem III.22, we deduce that the function $t \mapsto \langle g_t, \varphi_t\rangle$ is ν-measurable.

By theorem IV.16, the set

$$\Phi(t) = \{u \in S^{\cdot\cdot}_{\Gamma} \mid \int \langle g_t,u\rangle\, dm = \int \langle g_t, \varphi_t\rangle\, dm\}$$

is not empty for every $t \in T$. Since $S^{\cdot\cdot}_{\Gamma}$ is Polish and $(t,u) \mapsto \int \langle g_t,u\rangle\, dm$ is $\mathcal{T} \otimes \mathcal{B}(S^{\cdot\cdot}_{\Gamma})$-measurable (where $\mathcal{T}$ is the tribe of all ν-measurable sets of T), the multifunction Φ admits by theorem III.38 a measurable selection $\tilde{\varphi}$ with the required properties.

Appendix. Proof of Ljapunov's convexity theorem.

The proof of Ljapunov's convexity theorem that follows is due to Lindentrauss ([18]) and Arkin-Levin ([2]).

Theorem. Let $(\Omega, \mathcal{G}, \mu)$ be a measure space with finite positive non atomic measure μ. Let f_i $(i=1,2,\ldots,n)$ be a finite sequence of μ integrable real functions on Ω. Let U be the positive part of the unit ball in $L^{\infty}_{\mathbb{R}}(\Omega, \mathcal{G}, \mu)$, and $\ddot{U}$ the set of extreme points of U,

Then

$$\{\int \alpha f_i \, d\mu \mid \alpha \in \ddot{U}\} = \{\int \alpha f_i \, d\mu \mid \alpha \in U\}, \; i=1,\dots,n \, .$$

This equality implies that the set of values of the vector measure

$m = (f_1 \mu, \dots, f_n \mu)$

$$\{m(A) \mid A \in \mathfrak{G}\}$$

is convex and compact in $\mathbb{R}^n$.

The idea of the proof is due to Karlin ([14] and Karlin-Studden ([15]) as we have already seen in the remark concerning the proof of theorem IV.17 (See also the second proof of this theorem). Hence, by applying the techniques of the proof of theorem IV.17 and a very short construction due to Lindentrauss ([18]) of Arkin-Levin ([2]) to the special situation of the preceding theorem, we obtain Ljapunov's convexity theorem.rem.

Proof. Since the unit ball of $L^\infty_{\mathbb{R}}(\Omega, \mathfrak{G}, \mu)$ is $\sigma(L^\infty_{\mathbb{R}}(\Omega, \mathfrak{G}, \mu), L^1_{\mathbb{R}}(\Omega, \mathfrak{G}, \mu))$ compact and convex and the linear mapping

$$T : \alpha \mapsto (\int \alpha f_i \, d\mu)_{1 \leq i \leq n}$$

from $L^\infty_{\mathbb{R}}(\Omega, \mathfrak{G}, \mu)$ to $\mathbb{R}^n$ is continuous with respect to the topology $\sigma(L^\infty_{\mathbb{R}}, L^1_{\mathbb{R}})$, the set $T(U)$ is convex compact in $\mathbb{R}^n$. Let $\alpha_0 \in U$ and consider

$$T^{-1}(\alpha_0) = \{\alpha \in U \mid \int \alpha f_i \, d\mu = \int \alpha_0 f_i \, d\mu, \; i=1,\dots,n\}$$

Then $T^{-1}(\alpha_0)$ is convex and $\sigma(L^\infty_{\mathbb{R}}, L^1_{\mathbb{R}})$ compact; to prove the theorem, it suffices to show that the extreme points of $T^{-1}(\alpha_0)$ belongs to $\ddot{U}$. Let u_0 be an extremal point of $T^{-1}(\alpha_0)$ and suppose $u_0 \notin \ddot{U}$. Then there exists a strictly positive number η such that the $\mathfrak{G}$-measurable set

$$A = \{\omega \in \Omega \mid \eta \leq u_0(\omega) \leq 1-\eta\}$$

is non null(i.e. $\mu(A) > 0$). Since μ is non-atomic, the space $L^1_{\mathbb{R}}(A, \mathfrak{G}_A, \mu)$ is infinite-dimensional. Then, by the Hahn-Banach theorem, there exists an element v in the dual $L^\infty_{\mathbb{R}}(A, \mathfrak{G}_A, \mu)$ of $L^1_{\mathbb{R}}(A, \mathfrak{G}_A, \mu)$ such that

$$\begin{cases} v \neq 0 \\ |v(\omega)| \leq \eta \quad \text{for all} \quad \omega \in A \\ \int_A v\, f_i \, d\mu = 0 \ , \ i=1,2,\ldots,n. \end{cases}$$

Define

$$\tilde{u}(\omega) = \begin{cases} v(\omega) & \text{for} \quad \omega \in A \\ 0 & \text{for} \quad \omega \in \Omega \setminus A \end{cases}$$

Then $u_o \pm \tilde{u} \in T^{-1}(f_o)$ and

$$u_o(\omega) = \frac{u_o(\omega) + \tilde{u}(\omega)}{2} + \frac{u_o(\omega) - \tilde{u}(\omega)}{2} , \quad \forall\ \omega \in \Omega.$$

This contradicts the extremal property of u_o and achieves the proof.

Remark. The construction of $\tilde{u}$ in the proof of the preceding theorem is due to Arkin-Levin ([2]). The above proof is due to Lindentrauss ([18] who gives a different construction of $\tilde{u}$ by induction on the dimension of the space $\mathbb{R}^n$.

BIBLIOGRAPHY OF CHAPTER IV

1 - ALFSEN, E.M. - Compact convex sets and boundary integrals. Springer-Verlag. Berlin Heidelberg New York 1971.

2 - ARKIN, V.I. - LEVIN, V.L. - Convexity of values of vector integrals theorems on measurable choices and variationals problems. Russian Math. Surveys 27, 21-85 (1972).

3 - BENAMARA, M. -

[3.1] Sections extrémales d'une multi-application. C.R. Acad. Sc. Paris 278, 1249-1252 (1974).

[3.2] Points extrémaux multi-applications et fonctionnelles intégrales. Thèse de 3° cycle, Grenoble (1975).

4 - BERGE, C. - Espaces topologiques. Fonctions multivoques. Dunod 1958.

5 - BLACKWELL, D. - The range of certain vector integrals. Proc. Amer. Math. Society 2, 390-395 (1951).

6 - BOURBAKI, N. -

[6.1] Topologie générale. Chapitre 9. 2ème édition. Hermann Paris 1958.

[6.2] Intégration sur les espaces topologiques séparés. Chapitre 9 Hermann Paris 1969.

7 - CASTAING, Ch. -

[7.1] Sur les multi-applications mesurables. Revue d'Informatique et de Recherche opérationnelle 1, 91-126 (1967).

[7.2] Sur la mesurabilité du profil d'un ensemble convexe compact variant de façon mesurable. Collège Scientifique Universitaire Perpignan (1968). Polycopié.

[7.3] Sur une nouvelle extension du théorème de Ljapunov. C.R. Acad. Sc. Paris 264, 333-336 (1967).

[7.4] Application d'un théorème de compacité à un résultat de désintégration des mesures. C.R. Acad. Sc. Paris 270, 1732-1735 (1970).

8 - CHOQUET, G. - MEYER, P.A. - Existence et unicité des représentations intégrales dans les convexes compacts quelconques. Ann. Inst. Fourier (Grenoble) 13, 139-154 (1963).

9 - DUNFORD, N. - SCHWARTZ, J.T. - Linear operators. Part. I. Interscience 1964.

10 - DVORETZKY, A. - WALD, A. - WOLFOWITZ, J. - Relations among certain ranges of vector measures. Pacific J. Math. 1, 59-74 (1951).

11 - GODET-THOBIE, C. - Sur les multimesures de transition. C.R. Acad. Sc. Paris 278, 1367-1369 (1974) et exposé n° 5, Séminaire d'Analyse convexe Montpellier (1974).

12 - GHOUILA HOURI, A. - Sur la généralisation de la notion de commande d'un système guidable. Revue d'Informatique et de recherche opérationnelle 4, 7-32 (1967).

13 - HALMOS, P.R. - The range of a vector measure. Bull. Amer. Math. Soc. 54, 416-421 (1948).

14 - KARLIN, S. - On extreme points of vector functions. Proc. Amer. Math. Society 4, 603-610 (1953).

15 - KARLIN, S. - STUDDEN, W.J. - Tchebycheff systems ; with applications in Analysis and Statistics. John Wiley 1966.

16 - KELLERER, H.G. - Bemerkung zn einen satz von H. Richter. Archiv. der Math. 15, 204-207 (1964).

17 - KINGMAN, J.F.C. - ROBERTSON, A.P. - On a theorem of Liapounoff, J. of London Math. Soc. 43, 347-351 (1968).

18 - LINDENSTRAUSS, J. - A shorf proof of Liapounoff's convexity theorem. J. of Math. and Mechanics 15, 971-972 (1966).

19 - LJAPUNOV, A. - Sur les fonctions-vecteurs complètement additives. Bull. Acad. Sci. URSS sér. Math. 4, 465-478 (1940).

20 - MEYER, P.A. - Probabilités et Potentiel. Hermann 1966.

21 - NEVEU, J. - Bases mathématiques du calcul des probabilités. Masson 1964.

22 - OLECH, C. - Extremal solutions of a control system. J. Differential equations 2, 74-101 (1966).

23 - SAINTE-BEUVE, M.F. - On the extension of Von Neumann-Aumann's theorem. J. Functional Analysis 17, 112-129 (1974).

24 - STRASSEN, V. - The existence of probability measures with given marginals. Ann. Math. Stat. 36, 423-439 (1965).

25 - VALADIER, M. - Contribution à l'analyse convexe. Thèse Faculté des Sciences (Paris) (1970).

26 - WARGA, J. - Optimal control of Differential and Functional Equations, Academic Press, New York 1972.

CHAPTER V

COMPACTNESS THEOREMS OF MEASURABLE SELECTIONS AND INTEGRAL REPRESENTATION THEOREM.

§ 1 - COMPACTNESS THEOREMS IN THE SPACES $L^{\infty}_{E'}(\Omega,\mathcal{A},\mu)$ and $L^{1}_{E}(\Omega,\mathcal{A},\mu)$.

Let $(\Omega,\mathcal{A},\mu)$ be a σ-finite positive measure space with $\mathcal{A}$-μ complete and F a Suslin locally convex space. If Γ is a scalarly $\mathcal{A}$- measurable multifunction from Ω with non empty convex compact values in F, i.e, for every $x'\in F'$, the function $\omega' \to \delta^*(x', \Gamma(\omega))$ is $\mathcal{A}$-measurable on Ω, we denote by S_{Γ} the quotient of the set $\mathcal{S}_{\Gamma}$ of all $\mathcal{A}$-measurable selections of Γ for the equivalence "equality almost everywhere". By theorem III-37, we know that $\mathcal{S}_{\Gamma}$ is not empty ; if Γ is scalarly integrable, i.e., for every x' belonging to the dual of F, the scalar function $\omega \to \delta^*(x', \Gamma(\omega))$ is $\mathcal{A}$-measurable and μ-integrable, we show, in this section, the compactness of S_{Γ} for some adequate topologies, which thus have applications in the existence theory for various optimization problems, the theory of multivalued differential equations and the theory of Integration of multifunctions.

1 - Theorem V-1 - Let F be the weak dual E'_s of a separable Banach space E and let Γ be a scalarly measurable multifunction from Ω with non empty convex compact values in F. We suppose that $\Gamma(\omega)$ is contained in unit ball B' of E' for every $\omega \in \Omega$. Then, S_{Γ} is convex, compact for the weak topology $\sigma(L^{\infty}_{E'_s}, L^{1}_{E})$.

Proof. It is well known that the topological dual of the Banach space of integrable functions from Ω to E, L^{1}_{E}, is $L^{\infty}_{E'_s}$ (a short proof of this result is given by Meyer ([12], p. 301) ; (see also A. and C. Tulcea ([18])). Therefore, S_Γ is relatively compact for the weak topology $\sigma(L^\infty_{E'_s}, L^1_E)$. Let $\mathcal{U}$ be an ultrafilter on S_Γ and $\tilde{\dot{u}}$ be the limit of $\mathcal{U}$; then we have

$$\lim_{\dot{u} \in \mathcal{U}} < \dot{u}, f > = < \tilde{\dot{u}}, f >, \quad \forall f \in L^1_E$$

We must prove that $\tilde{\dot{u}}$ belongs to S_Γ. Suppose that $\tilde{\dot{u}} \notin S_\Gamma$. Then, by lemma III-34 and lemma III-32, we remark, first, that the following set

$$A = \{\omega \in \Omega |\ \tilde{u}(\omega) \notin \Gamma(\omega)\}$$

is $\mathcal{A}$-measurable because, there exists a sequence $(e_n)_{n \in \mathbb{N}} \in E$ such that

$$A = \bigcup_{n \in \mathbb{N}} \{\omega \in \Omega |\ < e_n, \tilde{u}(\omega) >> \delta^*(e_n, \Gamma(\omega))\}$$

So, there exists an $e_n \in E$ such that the following $\mathcal{A}$-measurable set

$$A_n = \{\omega \in \Omega |\ < e_n, \tilde{u}(\omega) > > \delta^*(e_n, \Gamma(\omega))\}$$

satisfies $0 < \mu(A_n)$. Since $\mathcal{U}$ converges weakly to $\tilde{\dot{u}}$, we have, by integrating on A_n,

$$\int_{A_n} < \tilde{u}, e_n > d\mu = \lim_{\dot{u} \in \mathcal{U}} \int_{A_n} < u, e_n > d\mu \leq \int_{A_n} \delta^*(e_n, \Gamma(\omega))\, d\mu$$

This inequality contradicts the following:

$$\int_{A_n} < \tilde{u}, e_n > d\mu > \int_{A_n} \delta^*(e_n, \Gamma(\omega))\, d\mu.$$

and completes the proof of the theorem.

2 - One of the important applications of the theorem V-1 is the following one which comes from the theory of relaxed control and from which we can derive other compactness results.

<u>Theorem</u> V-2 - <u>Let</u> U <u>be a compact metrizable space and let</u> Γ <u>be a measurable multifunction from</u> Ω <u>with non empty compact values in</u> U. <u>Let</u> $\mathcal{C}(U)$ <u>be the Banach space of all real continuous functions defined on</u> U,

equipped with the tobology of uniform convergence and let $\mathcal{M}(U)$ be the space of Radon measures defined on U, equipped with the weak topology $\sigma(\mathcal{M}(U), \mathcal{C}(U))$. Let $\mathcal{M}_+^1(U)$ be the set of positive Radon measures on U with mass one and let us put;

$$\Sigma(\omega) = \{\nu \in \mathcal{M}_+^1(U) \mid \nu(\Gamma(\omega)) = 1\}, \ \forall\, \omega \in \Omega$$

Then, the multifunction Σ with non empty convex compact values in $\mathcal{M}_+^1(U)$ is scalarly measurable and the quotient S_Σ of the set $\mathcal{S}_\Sigma$ of all measurable selections of Σ is not empty and compact for the weak topology

$$\sigma(L^\infty_{\mathcal{M}(U)}(\Omega, \mathcal{A}, \mu), L^1_{\mathcal{C}(U)}(\Omega, \mathcal{A}, \mu)).$$

Proof. Since U is a compact metrizable space, $\mathcal{C}(U)$ is a separable Banach space. On the other hand, the multifunction Σ is scalarly measurable. Indeed, by theorem III-8, Γ admits a countably dense $\mathcal{A}$-measurable selection $(u_n)_{n \in \mathbb{N}}$, so that, we have for every $f \in \mathcal{C}(U)$,

$$\delta^*(f, \Sigma(\omega)) = \sup_{n \in \mathbb{N}} f(u_n(\omega)), \quad \forall\, \omega \in \Omega.$$

Now, to finish the proof, it suffices to apply theorem V-1 to the locally convex space $F = \mathcal{M}(U)$ and the multifunction Σ.

3 - We give now another compactness result which is very useful in optimization problems and the theory of multivalued differential equations.

Theorem V-3 - Under hypotheses and notations of theorem V-2, let h be a mapping of $\Omega \times U$ into a separable Banach space G and let α be a positive μ-integrable function defined on Ω. Suppose that mapping h satisfies the following conditions :

a) for every $\omega \in \Omega$, $h(\omega, .)$ is continuous from U to G with respect to the weak topology $\sigma(G, G')$.

b) *for every* $u \in U$, $h(.,u)$ *is* $\mathcal{A}$*-measurable on* Ω

c) *for every* $\omega \in \Omega$ *and* $u \in U$, $\|h(\omega,u)\| \leq a(\omega)$.

Let us denote by S_Δ *the quotient of the set* $\mathcal{S}_\Delta$ *of all measurable selections of the multifunction*

$$\Delta(\omega) = \overline{co}(h(\omega, \Gamma(\omega))) \quad (\omega \in \Omega)$$

where $\overline{co}(h(\omega, \Gamma(\omega)))$ *is the closed convex hull of the* $\sigma(G, G')$ *compact set* $h(\omega, \Gamma(\omega))$. *Then,* S_Δ *is not empty and compact for the weak topology*

$$\sigma(L^1_G(\Omega, \mathcal{A}, \mu), L^\infty_{G'_s}(\Omega, \mathcal{A}, \mu)).$$

Proof. For every $\omega \in \Omega$ and $\lambda \in \mathcal{S}_{\Sigma'}$, let us put

$$A_\lambda(\omega) = \int_U h(\omega,u)\, d\lambda_\omega(u)$$

It is clear that $A_\lambda(\omega)$ belongs to G and the mapping $A_\lambda : \omega \to A_\lambda(\omega)$ of Ω into G is $\mathcal{A}$-measurable. Indeed, since G is separable, it is enough to check that this mapping is scalarly $\mathcal{A}$-measurable.

Let x' belong to the dual G' of G. Then the mapping $\rho_{x'}$: $\omega \to x' \circ h(\omega,.)$ of Ω into $\mathcal{C}(U)$ belongs to $L^1_{\mathcal{C}(U)}(\Omega, \mathcal{A}, \mu)$ since h satisfies conditions a), b), and c). On the other hand, the mapping λ belongs to $L^\infty_{\mathcal{M}(U)}(\Omega, \mathcal{A}, \mu)$, therefore, the scalar function

$$\omega \to \langle x', A_\lambda(\omega) \rangle = \int_U \langle x', h(\omega,u) \rangle d\lambda_\omega(u) = \langle \rho_{x'}(\omega), \lambda_\omega \rangle$$

is $\mathcal{A}$-measurable.

Now, let $g \in L^\infty_{G'_s}(\Omega, \mathcal{A}, \mu)$. Then the scalar function $\omega \to \langle g(\omega), h(\omega, u) \rangle$ is $\mathcal{A}$-measurable for every $u \in U$. By condition c) we have

$$\|A_\lambda(\omega)\| \leq a(\omega) \quad , \quad \forall\, \omega \in \Omega$$

$$|\langle g(\omega), h(\omega, u) \rangle| \leq a(\omega)\|g(\omega)\|, \qquad \forall \omega \in \Omega,\ \forall u \in U,$$

so, this implies

$$\langle A_\lambda, g \rangle = \int_\Omega d\mu(\omega) \int_U \langle g(\omega), h(\omega,u) \rangle\, d\lambda_\omega(u)$$

Denote by ρ_g the mapping $\omega \to \langle g(\omega), h(\omega,.)\rangle$ from Ω into $\mathcal{C}(U)$. Then ρ_g belongs to $L^1_{\mathcal{C}(U)}(\Omega, \mathcal{A}, \mu)$. Hence, by virtue of the above argument we use to prove the measurability of A_λ, we can write

$$\langle A_\lambda, g\rangle = \langle \rho_g, \lambda\rangle$$

where the brackets $\langle .,.\rangle$ denote the duality between $L^1_G(\Omega, \mathcal{A}, \mu)$ and $L^\infty_{G'_s}(\Omega, \mathcal{A}, \mu)$ and the duality between $L^\infty_{\mathcal{M}(U)}(\Omega, \mathcal{A}, \mu)$ and $L^1_{\mathcal{C}(U)}(\Omega, \mathcal{A}, \mu)$. It follows immediately that affine mapping $\lambda \to A_\lambda$ of S_Σ into $L^1_G(\Omega, \mathcal{A}, \mu)$ is continuous for the weak topology of $L^1_G(\Omega, \mathcal{A}, \mu)$. So, the image of S_Σ by the mapping $\lambda \to A_\lambda$ is convex and compact for the weak topology $\sigma(L^1_G(\Omega, \mathcal{A}, \mu), L^\infty_{G'_s}(\Omega, \mathcal{A}, \mu))$.

To finish the proof we assert, now, the equality

$$\{A_\lambda \mid \lambda \in S_\Sigma\} = S_\Delta$$

By a well known result ([1], Int. ch IV, 2^e ed.) we have already the inclusion

$$\{A_\lambda \mid \lambda \in S_\Sigma\} \subset S_\Delta$$

To prove the inverse inclusion, let us put

$$\varphi(\omega, \mu) = \int_U h(\omega, u)\, d\mu(\omega), \quad \forall\, \omega \in \Omega, \quad \forall\, \mu \in \mathcal{M}^1_+(U)$$

Then, it is obvious that $\varphi(\omega, .)$ is continuous on $\mathcal{M}^1_+(U)$ for every $\omega \in \Omega$ and $\varphi(., \mu)$ is $\mathcal{A}$-measurable for every $\mu \in \mathcal{M}^1_+(U)$. Let v be an element of $\mathcal{S}_\Delta$. Then, we have, **for ω almost everywhere,**

$$v(\omega) \in \varphi(\omega, \Sigma(\omega)).$$

Since Σ is a $\mathcal{A}$-measurable multifunction, we can apply the measurable implicit function theorem III-38 to obtain a $\mathcal{A}$-measurable selection λ of Σ such that

$$v(\omega) = \varphi(\omega, \lambda(\omega)) = \int_U h(\omega, u)\, d\lambda_\omega(\omega)$$

for μ a.e. ; this proves our assertion.

4 - The following corollary is very important for many applications that we have in mind (see also, for instance, Bismut ([2]) for another application of this corollary) and generalizes the following well known result : Given $\alpha \in L^1_+(\Omega, \mathcal{A}, \mu)$, then the set

$$\{f \in L^1_{\mathbb{R}}(\Omega, \mathcal{A}, \mu) \mid |f| \leq \alpha\}$$

is convex and compact for the weak topology $\sigma(L^1_{\mathbb{R}}(\Omega, \mathcal{A}, \mu), L^\infty_{\mathbb{R}}(\Omega, \mathcal{A}, \mu))$.

Corollary V-4. Let Φ be a $\mathcal{A}$-measurable multifunction from Ω with non empty convex $\sigma(E,E')$ compact values in a separable Banach space E, α a real positive function of $\mathcal{L}^1_R(\Omega, \mathcal{A}, \mu)$ and K a convex $\sigma(E,E')$ compact subset of E. Suppose that $\Phi(\omega) \subset \alpha(\omega) K$ for every $\omega \in \Omega$. Then, the quotient S_Φ of the set of all $\mathcal{A}$-measurable selections of Φ is convex and compact for the weak topology $\sigma(L^1_E(\Omega, \mathcal{A}, \mu), L^\infty_{E'_s}(\Omega, \mathcal{A}, \mu))$.

Proof. It is easy to verify that S_Φ is closed for the topology defined by the norm $N_1(f)$. Let (s_n) be a sequence in $\mathcal{S}_\Phi$ and $s \in \mathcal{L}^1_E(\Omega, \mathcal{A}, \mu)$ such that

$$\lim_{n \to \infty} N_1(s - s_n) = 0$$

then there exists a subsequence (s_{n_k}) such that (s_{n_k}) converges to s a. e. ; since $\Phi(\omega)$ is closed ; $s(\omega) \in \Phi(\omega)$ a.e. ; this proves our assertion. To finish the proof, it is enough to show that the quotient S_Δ of all $\mathcal{A}$-measurable selections of the $\mathcal{A}$-measurable multifunction Δ from Ω with convex $\sigma(E,E')$ compact values in E, $\omega \to \Delta(\omega) = \alpha(\omega) K$, is compact for the weak topology $\sigma(L^1_E(\Omega, \mathcal{A}, \mu), L^\infty_{E'_s}(\Omega, \mathcal{A}, \mu))$. For this purpose we can suppose without loss of generality that K is contained in the unit ball of E. Now, let us observe that K is a compact metrizable space for the weak topology $\sigma(E, E')$ since E is separable. Therefore we can apply theorem V-3 and its notations by taking

$$\begin{cases} h(\omega,u) = \alpha(\omega)u, & \omega \in \Omega, \quad u \in K = U \\ \Gamma(\omega) = K, & \omega \in \Omega \\ \Delta(\omega) = \alpha(\omega)\, K, & \omega \in \Omega \end{cases}$$

According to theorem V-3, S_Δ is convex, compact for the weak topology $\sigma(L^1_E, L^\infty_{E'_s})$, so is S_Φ because S_Φ is weakly closed and contained in S_Δ.

Remark 1) Corollary V-4 can also be derived, using special convexity arguments ([4]) we develop below. In addition, these arguments imply a variant of corollary V-4 when E is a reflexive Banach space, and, at the same time, some "inf-compactness" results.

2) When E is a reflexive separable Banach space, an inspection of the proof of theorem V-3 and corollary V-4 furnishes us with a weak compactness result in the theory of generalized Köthe function space. This has been shown recently in a different way([8]).

§ 2 - INF-COMPACTNESS THEOREMS.

5 - We first recall here a well known result whose proof we give here derives from ([6]) ; for a more general setting one can see Tulcea's book ([18]).

We shall denote in this paragraph by (T, μ) a locally compact space equipped with a positive Radon measure. Let E be a reflexive Banach space and E'_b the strong dual of E.

Theorem V-5 - There exists a linear isomorphic isometry between the strong dual $(L^1_E(T, \mu))'$ of the Banach space $L^1_E(T,\mu)$ and the Banach space $L^\infty_{E'_b}(T,\mu)$.

Proof. We shall use essentially the same method as in the proof of proposition 10 in Bourbaki ([1], Intégration, chap.6, p.47). Let us indicate first that the measurability we use here is given in the sense of Bourbaki.

Let u be a continuous linear mapping of $L^1_E(T,\mu)$ into $\mathbb{R}$. For every $h \in L^1(T,\mu)$, the mapping $x \to u(h\,x)$ is a continuous linear mapping of E into $\mathbb{R}$ since we have

$$|u(h\,x)| \le \|u\| N_1(h\,x) \le \|u\|.\|x\|.\,N_1(h)$$

Then, there exists an unique element $A(h)$ of E' such that

$$u(h\,x) = < x,\ A(h) >$$

But E is reflexive, then so is E'_b. Hence the linear mapping $h \to A(h)$ of $L^1_{\mathbb{R}}(T,\ \mu)$ into E'_b is weakly compact. By virtue of the Dunford-Pettis-Phillips theorem (Bourbaki ([1]), intégration, chap.6, p.95, ex. 24 ; see also Grothendieck, [10], theor. 3, p.314) there exists a μ-measurable mapping g of T into E'_b such that

$$\|g(t)\| \le \|u\| \qquad \text{a.e.}$$

$$u(h\,x) = < x,\ A(h) > = \int_T < hx,\ g > d\mu$$

for every $x \in E$ and every $h \in L^1_{\mathbb{R}}\ (T,\ \mu)$.

Let us consider now the following continuous linear mapping of $L^1_E(T,\mu)$ to $\mathbb{R}$.

$$\theta(g) \ :\ f \to \int < g,f > d\mu,\ \ f \in L^1_E(T,\ \mu)$$

(It is easily seen that $\theta(g)$ is continuous by using the Hölder inequality). Then, $\theta(g)$ and u are equal on the vector subspace of $L^1_E(T,\mu)$ generated by the elements hx ($h \in L^1_{\mathbb{R}}$, $x \in E$).

Since this subspace is strongly dense in the Banach space L^1_E, u is equal to θ (g). It follows immediately that $\theta(g) = N_\infty(g)$ where $N_\infty(g)$ is the norm of g, since, by using again the Hölder inequality, follows that

$$|\theta(g)| \le N_\infty(g) \le \|u\| = |\theta(g)|$$

6 - We recall the following lemma due to Grothendieck ([10])

<u>Lemma</u> V-6 - <u>Let</u> G <u>be a Banach space,</u> B <u>its unit ball. Let</u> H <u>be a subset of</u> G <u>which fulfills the following property : for every positive number</u> ε, <u>there exists a weakly compact set</u> K_ε <u>in</u> G <u>such that</u> $H \subset K_\varepsilon + \varepsilon B$. <u>Then,</u> H <u>is relatively weakly compact.</u>

<u>Proof.</u> The closure $\bar{H}$ of H in the bidual space G'' equipped with the weak topology $\sigma(G'', G')$ is compact since H is bounded. We have to show that $\bar{H}$ is contained in G. Let B'' be the unit ball of G''. Then $K_\varepsilon + \varepsilon B''$ is $\sigma(G'', G')$ closed since K and B'' are both $\sigma(G'',G')$ compact. Since we have by hypothesis $H \subset K_\varepsilon + \varepsilon B$, we deduce easily that $\bar{H}$ is contained in $K_\varepsilon + \varepsilon B''$, so that $\bar{H} \subset G + \varepsilon B''$. But ε is an arbitrary positive number and G is closed in G'' with respect to the norm topology of G'':, the last inclusion shows that $\bar{H} \subset G$.

7 - The following lemma is an inf-compactness result.

<u>Lemma V-7</u> - <u>Let</u> G <u>be a Banach space,</u> G' <u>its dual and</u> B' <u>the unit ball of</u> G'. <u>Let</u> $h \in \bar{\mathbb{R}}^{G'}$ <u>and</u> h^* <u>the polar function of</u> h, i.e.,

$$h^*(y) = \sup\{< x,y > - h(x) | x \in G'\} , \; y \in G$$

<u>Suppose that</u> h <u>satisfies the following condition :</u>

<u>For every number</u> $\varepsilon > 0$ <u>and every number</u> $r > 0$, <u>there exists a balanced convex</u> $\sigma(G,G')$ <u>compact set</u> H <u>such that</u>

$$x \in r B' \cap H^\circ \Rightarrow h(x) \leq \varepsilon$$

<u>where</u> H° <u>is the polar set of</u> H.

<u>Then,</u> h^* <u>is inf-</u>$\sigma(G,G')$ <u>compact, i.e., for every real number</u> ρ, <u>the following set</u> $h^* \leq (\rho)$

$$h^* \leq (\rho) = \{y \in G | h^*(y) \leq \rho\}$$

<u>is</u> $\sigma(G,G')$ <u>compact.</u>

<u>Proof</u>. It suffices to prove that, for every $\rho > 0$, $h^* \leq (\rho)$ is relatively $\sigma(G,G')$ compact. Let ε be a positive number. There exists, by hypothesis, a balanced convex $\sigma(G,G')$ compact set H of E such that

$$h \leq \varepsilon + \psi_{H^\circ \cap \frac{\rho+\varepsilon}{\varepsilon} B'}$$

where $\psi_{H^\circ \cap \frac{\rho+\varepsilon}{\varepsilon} B'}$ is the indicator function of the set $H^\circ \cap \frac{\rho+\varepsilon}{\varepsilon} B'$.

This implies immediately

$$h^* \geq - \varepsilon + \psi^*_{H^\circ \cap \frac{\rho+\varepsilon}{\varepsilon} B'}$$

where $\psi^*_{H^\circ \cap \frac{\rho+\varepsilon}{\varepsilon} B'}$ is the polar function of $\psi_{H^\circ \cap \frac{\rho+\varepsilon}{\varepsilon} B'}$

But the set $\{y \in G | \psi^*_{H^\circ \cap \frac{\rho+\varepsilon}{\varepsilon} B'}(y) \leq \rho + \varepsilon\}$ is homothetic to the set

$$\{y \in G | \psi^*_{H^\circ \cap \frac{\rho+\varepsilon}{\varepsilon} B'}(y) \leq 1\}$$

which is the polar set of $H^\circ \cap \frac{\rho+\varepsilon}{\varepsilon} B'$. By the classical formula of polarity, we have

$$(H^\circ \cap \frac{\rho+\varepsilon}{\varepsilon} B')^\circ = \overline{CO}(H^{\circ\circ} \cup (\frac{\rho+\varepsilon}{\varepsilon} B')^\circ) \subset H + \frac{\varepsilon}{\rho+\varepsilon} B$$

where $\overline{CO}(.)$ is the convex hull and $\dot{B}$ the unit ball in G. So, we deduce that

$$h^* \leq (\rho) \subset (\rho+\varepsilon) H + \varepsilon B$$

Since the set H is $\sigma(G,G')$ compact, by virtue of Grothendieck's lemma, $h^* \leq (\rho)$ is relatively $\sigma(G,G')$ compact.

8 - The following lemma is an inf-precompactness result.

<u>Lemma V-8</u> - <u>Let</u> G <u>be a locally convex space</u>, G' <u>its dual</u>. <u>Let</u> $h \in \overline{\mathbb{R}}^{G'}$ <u>and</u> h^* <u>the polar function of</u> h,

$$h^*(y) = \sup\{< x,y > - h(x) | x \in G'\} \ , \quad y \in G.$$

<u>Suppose that</u> h <u>satisfies the following condition</u> :

<u>For every</u> $\varepsilon > 0$ <u>and for every neighbourhood of origin in</u> G, V, <u>there exists a neighbourhood in</u> G' <u>for the weak topology</u> $\sigma(G',G)$, U', <u>such that</u>

$$h(x) \leq \varepsilon \ , \ \forall x \in V^{\circ} \cap U'$$

<u>where</u> V° <u>is the polar set of</u> V.

<u>Then</u>, h^* <u>is inf-precompact</u>, i.e., <u>for every real number</u> ρ, <u>the following set</u> $h^* \leq (\rho)$,

$$h^* \leq (\rho) = \{y \in G \mid h^*(y) \leq \rho\}$$

<u>is precompact in</u> G.

<u>Proof</u>. We shall use here the same techniques and notations as in the proof of lemma V-7.

Let ρ be a positive number. Let ε be a positive number and V a balanced convex closed neighbourhood of origin in G. There exists, by hypothesis, a neighbourhood U' of origin in G' for the weak topology $\sigma(G',G)$ such that

$$h \leq \varepsilon + \psi_{U' \cap (\rho+\varepsilon) V^{\circ}}$$

This implies

$$h^* \geq - \varepsilon + \psi^*_{U' \cap (\rho+\varepsilon) V^{\circ}}$$

whence $\quad h^* \leq (\rho) \subset \psi^*_{U' \cap (\rho+\varepsilon) V^{\circ}} \leq (\rho+\varepsilon)$

where $\psi^*_{U' \cap (\rho+\varepsilon) V^{\circ}} \leq (\rho+\varepsilon) = \{y \in G \mid \psi^*_{U' \cap (\rho+\varepsilon) V^{\circ}}(y) \leq \rho+\varepsilon\}$

But the last set is homothetic to the polar set of $U' \cap (\rho+\varepsilon) V^{\circ}$. So that, by using the well known formula of polarity, we have

$$(U' \cap (\rho+\varepsilon) V^{\circ})^{\circ} = \overline{co}(U'^{\circ} \cup (\rho+\varepsilon)^{-1} V) \subset U'^{\circ} + (\rho+\varepsilon)^{-1} V$$

Since U'° is compact in G, so is $(\rho+\varepsilon) U'^{\circ}$. Therefore, we can can find a finite sequence of elements in G', $x_1, x_2, \ldots, x_n$, such that

$$(\rho+\varepsilon)\, U'^{\circ} \subset \bigcup_{i=1}^{n} (x_i+V)$$

Finally we have

$$h^* \leq (\rho) \subset (\rho+\varepsilon)U'^{\circ} + V \subset \bigcup_{i=1}^{n} (x_i+V) + V \subset \bigcup_{i=1}^{n} (x_i+2V),$$

This proves that $h^* \leq (\rho)$ is **precompact.**

9 - Let $(\Omega, \mathcal{A}, \mu)$ be a σ-finite positive measure space with $\mathcal{A}_{-\mu}$ complete.

Theorem V-9 - Let E be a separable Banach space and K a balanced convex $\sigma(E,E')$ compact set of E. Let g be a mapping of $\Omega \times E'$ into $\mathbb{R}$. Assume the following assumptions :

(a) The mapping g is $\mathcal{A}\otimes\mathcal{B}(E'_s)$-measurable, where $\mathcal{B}(E'_s)$ is the Borel tribe of the weak dual E'_s of E,

(b) For every number $r > 0$, there exists a function, α_r, in $\mathcal{L}^1_{\mathbb{R}}(\Omega, \mathcal{A}, \mu)$ such that :

$$\forall\, \omega \in \Omega,\ \forall x' \in r\, B',\ |g(\omega,x')| \leq \alpha_r(\omega)\ \delta^*(x', K)$$

where B' is the unit ball in E'.

Let U' be the unit ball of $L^{\infty}_{E'_s}(\Omega, \mathcal{A}, \mu)$.

Then, the restriction to every subset $r\, U'$ $(r > 0)$ of the function

$$I_g : \dot{v} \to \int_{\Omega} g(\omega, v(\omega))\ \mu(d\omega),\ \dot{v} \in L^{\infty}_{E'_s}(\Omega, \mathcal{A}, \mu)$$

is continuous at the origin of $L^{\infty}_{E'_s}(\Omega, \mathcal{A}, \mu)$ with respect to the Mackey topology

$$\tau(L^{\infty}_{E'_s}(\Omega, \mathcal{A}, \mu),\ L^1_E(\Omega, \mathcal{A}, \mu)).$$

Proof. Let $(v_i,\ i \in D)$ be a generalized sequence in rU' $(r > 0)$ which converges to 0 with respect to the Mackey topology $\tau(L^{\infty}_{E'_s}, L^1_E)$. We have to show that the generalized sequence $(I_g(v_i))_{i \in D}$ converges to zero. Let α_r be a positive integrable function satisfying our assumptions. Since $(\Omega, \mathcal{A}, \mu)$ is σ-finite, we can find a countable partition $(\Omega_n)_{n \in \mathbb{N}}$ of Ω such that every Ω_n is $\mathcal{A}$-measurable with $\mu(\Omega_n) < +\infty$ and the restriction of α_r to each Ω_n is bounded.

Noting that the $\mathcal{A}$-measurable functions

$$\omega \to \alpha_r(\omega)\ \delta^*(v_i(\omega), K)$$

are majorized by a multiple of α_r since $x' \to \delta^*(x', K)$ is bounded on every bounded set of E', we write now

$$|I_g(v_i)| \le \sum_{n \le N_0} \int_{\Omega_n} \alpha_r(\omega)\ \delta^*(v_i(\omega), K)\ \mu\, d(\omega)$$
$$+ \sum_{n > N_0} \int_{\Omega_n} \alpha_r(\omega)\ \delta^*(v_i(\omega), K)\ \mu\,(d\omega)$$

Given a number $\varepsilon > 0$ we can choose N_0 such that the second integral is $\le \varepsilon$. If we denote by S_K the quotient of all $\mathcal{A}$-measurable selections of the constant multifunction, K, then $\chi_{\Omega_n} S_K$ (where χ_{Ω_n} is the characteristic function of Ω_n) is convex and $\sigma(L^1_E(\Omega_n), L^\infty_{E'_s}(\Omega_n))$ compact by virtue of corollary V-4. Furthermore, we can evaluate the support functional of $\chi_{\Omega_n} S_K$, by using selection theorem III-38,

$$\delta^*(\dot{v}, \chi_{\Omega_n} S_K) = \int_{\Omega_n} \delta^*(v(\omega), K)\ \mu(d\omega),\ \dot{v} \in L^\infty_{E'_s}(\Omega_n)$$

This formula implies

$$\lim_{i \in D} \int_{\Omega_n} \alpha_r(\omega)\ \delta^*(v_i(\omega), K)\ \mu(d\omega) = 0$$

because α_r is bounded on Ω_n and since $\dot{v} \to \delta(\dot{v}, \chi_{\Omega_n} S_K)$ is the support functional of the convex $\sigma(L^1_E(\Omega_n), L^\infty_{E'_s}(\Omega_n))$ compact set, $\chi_{\Omega_n} S_K$. So, it follows immediately that

$$\lim_{i \in D} \sum_{n \le N_0} \int_{\Omega_n} \alpha_r(\omega)\ \delta^*(v_i(\omega), K)\ \mu(d\omega) = 0$$

That achieves the proof.

10 - <u>Corollary V-10</u> - <u>Let</u> f <u>be a mapping of</u> $\Omega \times E$ <u>into</u> $]-\infty, +\infty]$. <u>Assume the following assumptions</u> :

(a) The mapping f is $\mathcal{A} \otimes \mathcal{B}(E)$-measurable, where $\mathcal{B}(E)$ is the Borel tribe of E,

(b) For every $\dot{u} \in L^1_E(\Omega, \mathcal{A}, \mu)$, set

$$I_f(\dot{u}) = \begin{cases} \int_\Omega f(\omega, u(\omega))\, \mu(d\omega) & \text{if } \int_\Omega f(\omega, u(\omega))^+ \mu(d\omega) < +\infty \\ +\infty & \text{otherwise} \end{cases}$$

If $I_f(\dot{u}) \geq \sup \{< \dot{u}, v > - I_g(\dot{v}) \mid v \in L^\infty_{E'_s}(\Omega, \mathcal{A}, \mu)\}$ for every $\dot{u} \in L^1_E(\Omega, \mathcal{A}, \mu)$, then, for every $\rho \in \mathbb{R}$,

$$I_f \leq (\rho) = \{\dot{u} \in L^1_E(\Omega, \mathcal{A}, \mu) \mid I_f(\dot{u}) \leq \rho\}$$

is relatively $\sigma(L^1_E, L^\infty_{E'_s})$ compact.

Proof. Let $(I_g)^*$ be the popular function of I_g

$$(I_g)^*(\dot{u}) = \sup \{< \dot{u}, \dot{v} > - I_g(\dot{v}) \mid \dot{v} \in L^\infty_{E'_s}\}$$

Then, by combining lemma V-7 and theorem V-9, we see that $(I_g)^*$ is inf-compact for the $\sigma(L^1_E, L^\infty_{E'_s})$ topology ; let $\rho \in \mathbb{R}$, then we have

$$(I_f) \leq (\rho) \subset (I_g)^* \leq (\rho)$$

since by (c), $I_f(\dot{u}) \geq (I_g)^*(\dot{u})$. So, $I_f \leq (\rho)$ is relatively $\sigma(L^1_E, L^\infty_{E'_s})$ compact.

Remarks 1) In chapter VII we shall describe $(I_g)^*$ in an integral form if g satisfies some rather general hypothesis.

2) Let us come back to the situation of theorem V-9. If we suppose that (T, μ) is σ-finite and if E is a reflexive Banach space and K is the unit ball of E, then, using essentially the same method as in the proof of theorem V-9 we obtain the following variant of theorem V-9.

- Theorem V-11 - Let (T, μ) be a locally compact space with μ σ-finite equipped with a positive Radon measure μ. Let E be a reflexive Banach space and E'_b its strong dual. Let g be a mapping of $T \times E'$ into $\mathbb{R}$ satisfying the following assumptions :

(a) The mapping g is $\mathcal{T}\otimes\mathcal{B}(E'_b)$-measurable, where $\mathcal{T}$ is the tribe of μ-measurable sets in T and $\mathcal{B}(E'_b)$ is the Borel tribe of E'_b,

(b) For every number $r > 0$, there exists a positive μ-integrable function α_r of $\mathcal{L}^1_{\mathbb{R}}(T,\mu)$ such that

$$\forall t \in T,\ \forall x' \in rB',\ |g(t, x')| \le \alpha_r(t)\|x'\|.$$

Let U be the unit ball of $L^\infty_{E'_b}(T,\mu)$. Then the restriction to every subset $r\,U'(r > 0)$ of the function

$$I_g : v \to \int_T g(t,v(t))\mu(dt),\ v \in L^\infty_{E'_b}(T,\mu)$$

is continuous at the origin of $L^\infty_{E'_b}(T,\mu)$ with respect to the Mackey topology $\tau(L^\infty_{E'_b}(T,\mu),\ L^1_E(T,\mu))$.

Proof. We shall use the same notations as in the proof of theorem V-9. We first observe that, for every μ-measurable subset A of T, with $\mu(A) < +\infty$, the unit ball of $L^\infty_E(A,\mu)$ is $\sigma(L^1_E(A,\mu),\ L^\infty_{E'_b}(A,\mu))$ compact. Indeed, since $\mu(A)$ is bounded, the injection $L^\infty_E(A,\mu) \to L^1_E(A,\mu)$ is linear and continuous for the topology $\sigma(L^\infty_E(A,\mu),\ L^1_{E'_b}(A,\mu))$ and $\sigma(L^1_E(A,\mu),\ L^\infty_{E'_b}(A,\mu))$. Now, by taking for K in the proof of theorem V-9 the unit ball of E, we have

$$|I_g(v_i)| \le \sum_{n\le N_0} \int_{\Omega_n} \alpha_r(t)\|v_i(t)\|\mu(dt) + \sum_{n>N_0} \int_{\Omega_n} \alpha_r(t)\|v_i(t)\|\mu(dt)$$

Therefore, we can repeat word for word the arguments we use in the proof of of theorem V-9, by observing that, for every n, the generalized sequence (v_i) converges to zero with respect to the strong topology of $L^1_{E'_b}(\Omega_n,\mu)$, since the unit ball of $L^\infty_E(\Omega_n,\mu)$ is $\sigma(L^1_E(\Omega_n,\mu),\ L^\infty_{E'_b}(\Omega_n,\mu))$ compact as we have already observed. So, we conclude that

$$\lim_{i\in D'} I_g(v_i) = 0$$

when $(v_i, i \in D)$ converges to 0 on r U' with respect to the Mackey topology $\tau(L^{\infty}_{E'_b}, L^1_E)$.

Remarks 1) The reader can obtain, by combining lemma V-7 and theorem V-11 when E is a reflexive space, a variant of corollary V-10. For the connection with Rockafellar's [14],[15] and Valadier's [19] results we refer to Valadier's paper : there are some technical refinements of inf-compactness results in the paper of Valadier.

2) If T is a compact space, E is a reflexive Banach space, then we can easily obtain a compactness result in $L^1_E(T,\mu)$ by combining Grothendieck's lemma and theorem V-5 ; this fact was first used in Castaing's papers. Let H be a subset of $L^1_E(T,\mu)$ satisfying the following property :

$$(P) \quad \forall \epsilon > 0, \exists \eta > 0 \Rightarrow \int_{[|f|>\eta]} |f| \, d\mu \leq \epsilon, \quad \forall f \in H$$

where $\quad [|f| > \eta] = \{t \in T \mid \|f(t)\| > \eta\}$

Then H is relatively $\sigma(L^1_E(T,\mu), L^{\infty}_{E'_b}(T,\mu))$ compact. For every μ-measurable set Z, let us denote by χ_Z the characteristic function of Z. From the equality

$$f = \chi_{[|f| \leq \eta]} f + \chi_{[|f| > \eta]} f$$

it follows immediately from the property (P) that

$$H \subset K_\eta + \epsilon U$$

where K_η is the set of all μ-measurable mappings f of T into E such that $\|f(t)\| \leq \eta$ a.e. and U is unit ball of $L^1_E(T,\mu)$. Since K_η is $\sigma(L^1_E, L^{\infty}_E)$ compact, we conclude that H is relatively $\sigma(L^1_E, L^{\infty}_{E'_b})$ compact.

§ 3 - EXTREME POINTS OF CERTAIN CONVEX SETS OF MEASURABLE FUNCTIONS DEFINED BY UNILATERAL INTEGRAL CONSTRAINTS.

12 - The following theorem describes the extreme points of certain convex compact sets in $L^{\infty}_{E'_s}(\Omega, \mathcal{A}, \mu)$ where $(\Omega, \mathcal{A}, \mu)$ is a σ-finite measure space with $\mathcal{A}-\mu$ complete and E'_s is the weak dual of a separable Banach space E.

Theorem V-12 - Let Γ be a scalarly $\mathcal{A}$-measurable multifunction from Ω with non empty convex compact values in E'_s as in theorem V-1 and let S_Γ be the quotient of the set $\mathcal{S}_\Gamma$ of all scalarly $\mathcal{A}$-measurable selections of Γ. Assume that μ is non atomic. Let $f_i (i=1,2,..,n)$ be a sequence of mappings from $\Omega \times E'_s$ into $]-\infty, +\infty]$ satisfying the following conditions :

(i) The mappings f_i are $\mathcal{A} \otimes \mathcal{B}(E'_s)$-measurable, where $\mathcal{B}(E'_s)$ is the Borel tribe of E'_s

(ii) The mappings

$$u \to I_{f_i}(u) = \begin{cases} \int_\Omega f_i(\omega, u(\omega)) \ \mu(d\omega) & \text{if } \int_\Omega f_i(\omega, u(\omega))^+ \mu(d\omega) < +\infty \\ +\infty & \text{otherwise} \end{cases}$$

are lower-semi-continuous on $L^{\infty}_{E'_s}(\Omega, \mathcal{A}, \mu)$ with respect to the weak topology $\sigma(L^{\infty}_{E'_s}, L^1_E)$.

(iii) For all real numbers α and β, and u and v belonging to S_Γ,

$$I_{f_i}(\alpha u + \beta v) = \alpha I_{f_i}(u) + \beta I_{f_i}(v) \ ; \ i = 1,2,\ldots,n$$

Consider $\quad B = \{u \in S_\Gamma | I_{f_i}(u) \le 1, \ i = 1,2,\ldots,n\}$

Then

1) B is convex compact for the weak topology $\sigma(L^{\infty}_{E'_s}, L^1_E)$.

2) The profile $\ddot{B}$ (i.e. the set of all extreme points of B) is included in the profile $\ddot{S}_\Gamma (= S_{\ddot{\Gamma}})$ of S_Γ.

Proof. The proof is mutatis mutandis given in theorem IV-17.

First, it follows from conditions (i), (ii), (iii) that B is convex compact for the weak topology $\sigma(L^{\infty}_{E'_s}, L^1_E)$ since we have already shown by theorem V-1 that S_Γ is convex and compact for the weak topology $\sigma(L^\infty_{E'_s}, L^1_E)$. Second, let $u_0 \in \ddot{B}$ and assume that $u_0 \notin S_\Gamma^{\cdot\cdot}$. Hence, the set

$$A = \{\omega \in \Omega | u_0(\omega) \in \Gamma(\omega) \setminus \ddot{\Gamma}(\omega)\}$$

is $\mathcal{A}$-measurable and $\mu(A)$ is positive. By the construction given in theorem IV-17, there exists a $\mathcal{A}$-measurable mapping v from A into E'_s such that

$$\begin{cases} v(\omega) \neq 0, & \forall \omega \in A \\ u_0(\omega) \pm v(\omega) \in \Gamma(\omega), & \forall \omega \in \Omega \end{cases}$$

and we have

$$u_0(\omega) = \frac{u_0(\omega) + v(\omega)}{2} + \frac{u_0(\omega) - v(\omega)}{2}$$

It follows from the definition of B and the linearity of I_{f_i} $(i=1,2,\ldots,n)$ that

$$\int_A f_i(\omega, v(\omega))\, \mu(d\omega) < +\infty, \quad i = 1,2,\ldots,n.$$

Since μ is non atomic, we can now repeat the arguments given in theorem IV-17 to finish the proof.

Remark. Property 2) implies easily that $S_\Gamma^{\cdot\cdot}$ is dense in S_Γ for the weak topology $\sigma(L^\infty_{E'_s}, L^1_E)$. Moreover, one can adapt the proof of theorem V-12 to other analogous density theorems. See, for instance ([8]) a direct proof of the density theorem in generalized Köthe function space.

§ 4 - COMPACTNESS THEOREM IN GENERALIZED KÖTHE FUNCTIONS SPACES AND ITS APPLICATIONS.

13 - The following compactness result is very useful in the theory of Integration of multifunctions and the theory of multivalued differential equations

Let $(\Omega, \mathcal{A}, \mu)$ be a σ-finite positive measure space with $\mathcal{A}$ μ-complete. A real valued $\mathcal{A}$-measurable function f on Ω is locally integrable if

$$\int_A |f|\, d\mu < +\infty$$

for each $\mathcal{A}$-measurable subset A of Ω with $\mu(A) < +\infty$. For a given subset $\mathcal{G}$ of the space $\mathcal{L}$ of all real valued locally integrable functions, we define L and L^* to be the following subspaces of $\mathcal{L}$:

$$L = \{f \in \mathcal{L} \mid \int_\Omega |fg| d\mu < +\infty , \quad \forall g \in \mathcal{G}\}$$

$$L^* = \{h \in \mathcal{L} \mid \int_\Omega |fh| d\mu < +\infty , \quad \forall f \in L\}$$

Let Λ(resp. Λ^*) be the quotient for the equivalence relation "equality almost everywhere" of the vector space L(resp. L^*).

Λ is called a Köthe function space and Λ^* is the Köthe dual of Λ.

The vector spaces Λ and Λ^* are placed in duality by the bilinear form

$$< f,g > = \int_\Omega fg \, d\mu, \quad f \in \Lambda , \ g \in \Lambda^*$$

Moreover, Λ and Λ^* are order complete vector lattices. If E is a locally convex space, we denote by Λ_E the quotient for the equivalence relation "equality almost everywhere" of the vector space of all scalarly $\mathcal{A}$-measurable mappings from Ω to E such that for every $x' \in E'$, $< x',f > \in \Lambda$; Λ_E is called <u>a generalized Köthe function space</u>.

<u>Theorem V-13</u> - <u>Let</u> Γ <u>be a scalarly</u> $\mathcal{A}$-<u>measurable multifunction from</u> Ω <u>with non empty convex compact values in a locally convex Suslin space</u> E. <u>If, for every</u> $x' \in E'$, <u>the function</u> $\delta^*(x',\Gamma(.))$ <u>belongs to</u> Λ, <u>for every</u> $g \in \Lambda^*$ <u>and every scalarly</u> $\mathcal{A}$-<u>measurable selection</u> f <u>of</u> Γ, <u>the weak integral of the scalarly integrable function</u> gf <u>belongs to</u> E, <u>then the quotient</u> S_Γ <u>of the set</u> $\mathcal{S}_\Gamma$ <u>of all scalarly</u> $\mathcal{A}$-<u>measurable selections of</u> Γ <u>is a non empty convex compact subset of</u> Λ_E <u>for the topology of convergence on</u> $\Lambda^* \otimes E'$; <u>namely the topology</u> $\sigma(L_E, \Lambda^* \otimes E')$

<u>Proof.</u> We use here some arguments given by Valadier (20) in the case where Λ is $L^1(\Omega,\mathcal{A},\mu)$. Define by $\mathcal{H}$ the set of all linear mappings ϕ from E' into Λ such that

$$\forall\, x' \in E' , \quad \phi(x') \le \delta^*(x', \Gamma(.)).$$

We first assert that there exists a bijective mapping from S_Γ onto $\mathcal{H}$. Indeed, let σ be an element of $\mathcal{S}_\Gamma$. Then, the linear mapping $\Phi : x' \to < x', \sigma(.) >$ of E' into Λ belongs to $\mathcal{H}$ and $\sigma \to \Phi$ defines an injective mapping of S_Γ into $\mathcal{H}$. It remains to prove the more difficult fact that $\sigma \to \Phi$ is onto. Let τ be an arbitrary lift from Λ into L which is linear : such a lift τ is obtained by lifting in an arbitrary manner an algebraic basis of Λ. If Φ is given in $\mathcal{H}$, for each $\omega \in \Omega$ and each $x' \in E'$, let us set $g(\omega, x') = \tau[\Phi(x')]\,(\omega)$. Then for fixed x', $g(., x')$ belongs to L, and for fixed ω, $g(\omega, .)$ is a linear form on E'. So we can write $g(\omega, x') = < x', \sigma(\omega) >$; now, since E is Suslin, there exists a dense sequence (e'_n) in E' for the weak topology $\sigma(E', E)$. Let $\mathcal{F}$ be the countable set of linear combinations with rational coefficients of the elements e'_n and let F' be the vector space generated by the elements e'_n. Then $\mathcal{F}$ is dense in F' for any locally convex topology placed on F'. But F' is dense in E' for the weak topology $\sigma(E', E)$ as well as for the Mackey topology $\tau(E', E)$ by Mackey's theorem, so $\mathcal{F}$ is dense in E' for the Mackey topology $\tau(E', E)$. Let N be a negligible set such that

$$g(\omega, x') = < x', \sigma(\omega) > \le \delta^*(x', \Gamma(\omega))$$

for every $\omega \notin N$ and every $x' \in \mathcal{F}$. Hence, for every fixed $\omega \notin N$, $\sigma(\omega)|_{F'}$ can be extended to a unique $\tau(E', E)$ continuous linear form $s(\omega)$ because $\mathcal{F}$ is dense in E' for this topology. For every $\omega \in \Omega \setminus N$, let us put $\Gamma_{-1}(\omega) = \Gamma(\omega)$ and for $n \ge 0$

$$\Gamma_n(\omega) = \{x \in \Gamma_{n-1}(\omega) \mid < e'_n, x > = < e'_n, \sigma(\omega) >\}$$

This implies $\{s(\omega)\} = \bigcap_n \Gamma_n(\omega)$ for all $\omega \in \Omega \setminus N$. Put $s(\omega) = 0$ if

$\omega \in N$. Then s is an $\mathcal{A}$-measurable mapping from Ω to E by virtue of theorem III-23 such that $s(\omega) \in \Gamma(\omega)$ almost everywhere. Now, to finish the proof, it is enough to verify that for every $x' \in F'$, there exists a negligible set $N_{x'}$ such that $< x', s(\omega) > = < x', \sigma(\omega) >$ for $\omega \notin N_{x'}$. Indeed, for every $x' \in F'$ and every $g \in \Lambda^*$, we have

$$< x', \int g\,\sigma\, d\mu > = \int < x', g\sigma > d\mu = \int < x', gs > d\mu = < x', \int gs\, d\mu >$$

As $\int hs\, d\mu$ belongs to E for all $h \in \Lambda^*$ this implies that

$$< x', \sigma(\omega) > = < x', s(\omega) > \qquad \text{a.e.}$$

Now, the compactness of S_Γ is equivalent to the compactness of $\mathcal{H}$ in $\Lambda^{E'}$ when Λ is equipped with the weak topology $\sigma(\Lambda, \Lambda^*)$.

For every $x' \in E'$ and every $\varphi \in \mathcal{H}$, we have

$$-\delta^*(-x', \Gamma(.)) \leq \varphi(x') \leq \delta^*(x', \Gamma(.))$$

This implies that $\varphi(x')$ belongs to the interval (for the natural order of Λ)

$$[-\delta^*(-x', \Gamma(.)), \delta^*(x', \Gamma(.))]$$

which is compact for the weak topology $\sigma(\Lambda, \Lambda^*)$ (See the following remark) Let $\mathcal{U}$ be an ultrafilter on $\mathcal{H}$ which converges to $\tilde{\varphi}$. For every $x' \in E'$, we have

$$\lim_{\varphi \in \mathcal{U}} \varphi(x') = \tilde{\varphi}(x')$$

so that $\tilde{\varphi}$ is linear and $\tilde{\varphi}(x') \leq \delta^*(x', \Gamma(.))$. This proves that $\tilde{\varphi}$ belongs to $\mathcal{H}$ and completes the proof.

<u>Remark</u>. The compactness of an interval $[\alpha, \beta]$ in a Köthe space is well known and can be deduced easily from theorem V-3 and remark 2) of corollary V-4. Here we only sketch the proof of this property for the convenience of the reader, It is also <u>valid for vector Köthe function spaces</u> ([8]). First, with the notations of theorem V-3, assume $G = \mathbb{R}$ and α belongs to Λ, then we assert that S_Δ is not empty and compact

for the topology $\sigma(\Lambda,\Lambda^*)$. Indeed, by using the notations of the proof of theorem V-3, and the hypothesis, it is easily checked that, for every $\lambda \in S_\Sigma$, where $\Sigma(\omega) = \{\nu \in \mathcal{M}^1_+(U) | \nu(\Gamma(\omega)) = 1\}$ $(\omega \in \Omega)$, the mapping $\omega \to A_\lambda(\omega) = \int_U h(\omega,u)d\lambda_\omega(u)$ is $\mathcal{A}$-measurable and $|A_\lambda(\omega)| \le \alpha(\omega)$ for all $\omega \in \Omega$. This shows that $A_\lambda \in \Lambda$ for all $\lambda \in S_\Sigma$. If g is an element of Λ^*, then $|g(\omega)h(\omega,u)| \le \alpha(\omega)|g(\omega)|$ for all $\omega \in \Omega$. This implies

$$< A_\lambda, g > = \int_\Omega d\mu(\omega)\int_U g(\omega)h(\omega,u)d_{\lambda_\omega}(u)$$

Denote by ρ_g the mapping $\omega \to g(\omega)h(\omega,.)$ from Ω into $\mathcal{C}(U)$. Then ρ_g belongs to $L^1_{\mathcal{C}(U)}(\Omega,\mathcal{A},\mu)$. So we have

$$< A_\lambda, g > = < \rho_g, \lambda >$$

where the brackets denote the duality between Λ and Λ^* and the duality between $L^\infty_{\mathcal{M}(U)}(\Omega,\mathcal{A},\mu)$ and $L^1_{\mathcal{C}(U)}(\Omega,\mathcal{A},\mu)$. By using the same construction as in the proof of theorem V-3, we show easily that the following set

$$\{A_\lambda | \lambda \in S_\Sigma\} = S_\Delta$$

is compact for the topology $\sigma(\Lambda,\Lambda^*)$. Second, if we take

$$h(\omega,u) = \alpha(\omega)u, \quad \omega \in \Omega, \ u \in [-1,1]$$

$$\Gamma(\omega) = [-1,1] \quad \text{for all} \quad \omega \in \Omega$$

$$\Delta(\omega) = h(\omega,\Gamma(\omega)) = [-\alpha(\omega),\alpha(\omega)]$$

Then, clearly the interval $S_\Delta = [-\alpha,\alpha]$ is $\sigma(\Lambda,\Lambda^*)$ compact ; whence any interval of the form $[\alpha,\beta]$ is $\sigma(\Lambda, \Lambda^*)$ compact too.

14 - In the following, we give the compactness of the integral of a multifunction and a generalization of Strassen's theorem ([16]).

Theorem V-14 - With the hypotheses of theorem V-13, we define, for every $g \in \Lambda^*$, the weak integral $\int g\Gamma \, d\mu$ of $g\Gamma$ by

$$\int g\Gamma \, d\mu = \{\int gf \, d\mu | f \in S_\Gamma\}$$

where $\int gf \, d\mu$ is the weak integral of the scalarly integrable function gf. Then, $\int g\Gamma \, d\mu$ is a convex, $\sigma(E, E')$ compact set of E. In addition,

<u>the support functional of</u> $\int g\Gamma \, d\mu$ <u>is given by the formula</u>

$$\delta^*(x', \int g\Gamma \, d\mu) = \int_\Omega \delta^*(x', g(\omega)\Gamma(\omega))\mu(d\omega)$$

<u>for all</u> $x' \in E'$.

<u>Proof</u>. By theorem V-13, S_Γ is convex and compact for the topology of pointwise convergence on $\Lambda^* \otimes E'$. Whence $\int g\Gamma \, d\mu$ is convex and $\sigma(E,E')$ compact because $\int g\Gamma \, d\mu$ is the image of the continuous mapping $f \to \int gf \, d\mu$ from S_Γ to E when S_Γ is equipped with the topology of pointwise convergence on $\Lambda^* \otimes E'$ and E is equipped with the topology $\sigma(E,E')$ because we have

$$< x', \int gf \, d\mu > = < g \otimes x', f >$$

For every $x' \in E'$, we have

$$\forall f \in S_\Gamma , \quad < x', \int gf \, d\mu > \leq \int \delta^*(x', g(\omega)\Gamma(\omega))d\mu(\omega)$$

Consider the multifunction $\Delta_{x'}$

$$\Delta_{x'}(\omega) = \{x \in g(\omega)\Gamma(\omega) | \quad < x',x > = \delta^*(x', g(\omega)\Gamma(\omega))\} \quad (\omega \in \Omega)$$

Then, it is easily checked that the graph of $\Delta_{x'}$ belongs to $\mathcal{A} \otimes \mathcal{B}(E)$; consequently $\Delta_{x'}$ admits a $\mathcal{A}$-measurable selection (theorem III-37), i.e

$$\begin{cases} s(\omega) \in g(\omega)\Gamma(\omega) \\ < x', s(\omega) > = \delta^*(x', g(\omega)\Gamma(\omega)). \end{cases}$$

for all $\omega \in \Omega$. By theorem III-38, there exists $f \in \mathcal{S}_\Gamma$ such that $s(\omega) = g(\omega) f(\omega)$ for all $\omega \in \Omega$. That completes the proof.

<u>Remark</u>. If Γ is scalarly integrable, then theorem V-13 shows that S_Γ is compact for the topology of pointwise convergence on $L^\infty \otimes E'$ and the weak integral $\int \Gamma \, d\mu$ of the multifunction Γ defined by

$$\int \Gamma \, d\mu = \{\int f \, d\mu | f \in S_\Gamma\}$$

is convex and $\sigma(E,E')$ compact and we have

$$\delta^*(x', \int \Gamma \, d\mu) = \int_\Omega \delta^*(x', \Gamma(\omega))d\mu(\omega)$$

In the following theorem, we give sufficient conditions for which the weak integral $\int \Gamma \, d\mu$ is compact for the topology of E.

15 - Theorem V-15 - Let E be a quasi-complete(*) locally convex Suslin space and Γ a scalarly integrable multifunction defined on Ω with non empty convex compact values in E. Assume the following conditions :

(i) $\int \Gamma \, d\mu = \{\int f \, d\mu \mid f \in S_\Gamma\}$ is included in E

(ii) For every fixed equicontinous subset A in E' there exists a positive integrable function h_A such that

$$\sup_{x' \in A} |\delta^*(x', \Gamma(\omega))| \le h_A(\omega)$$

for all $\omega \in \Omega$.

Then, the weak integral $\int \Gamma \, d\mu$ is a convex compact set in E.

Proof. By theorem V-14, we know that $\int \Gamma \, d\mu$ is a convex, $\sigma(E,E')$ compact set in E and the support function of $\int \Gamma \, d\mu$ is defined by

$$\delta^*(x', \int \Gamma \, d\mu) = \int_\Omega \delta^*(x', \Gamma(\omega)) \, \mu(d\omega)$$

for all $x' \in E'$. Thus, it suffices to prove that $\int \Gamma \, d\mu$ is precompact because E is quasi-complete. Let ψ be the indicator function of $\int \Gamma \, d\mu$

$$\psi(x) = \begin{cases} 0 & \text{if } x \in \int \Gamma \, d\mu \\ +\infty & \text{if } x \notin \int \Gamma \, d\mu \end{cases}$$

It is clear that ψ is the polar function of the function $x' \to \delta^*(x', \int \Gamma d\mu)$. So, by virtue of lemma V-8, we have to prove that the restriction of the function $x' \to \delta^*(x', \int \Gamma \, d\mu)$ to any equicontinuous subset of E' is continuous with respect to the topology $\sigma(E',E)$. Now, given an equicontinuous subset A of E', we can suppose without loss of generality that

(*) A locally convex space is quasi-complete if every bounded and closed subset of this space is complete.

A is $\sigma(E',E)$ closed since the closed convex hull of A is still equicontinuous. Since A is a compact metrizable space when A is endowed with the weak topology $\sigma(E',E)$, it remains to prove that for every sequence (x'_n) in A converging to $x' \in A$ in this compact metrizable space, we have

$$\lim_{n \to \infty} \delta^*(x'_n, \int \Gamma \, d\mu) = \delta^*(x', \int \Gamma \, d\mu)$$

By Strassen's formula given in theorem V-14, we have to prove that

$$\lim_{n \to \infty} \int_\Omega \delta^*(x'_n, \Gamma(\omega))\mu(d\omega) = \int_\Omega \delta^*(x', \Gamma(\omega))\mu(d\omega)$$

As $\Gamma(\omega)$ is convex compact, $x' \to \delta^*(x', \Gamma(\omega))$ is $\sigma(E',E)$ continuous on A because the topology $\sigma(E',E)$ and the topology of uniform convergence on compact sets of E are equal on equicontinuous subsets of E'. Whence, for every fixed $\omega \in \Omega$, we have

$$\lim_{n \to \infty} \delta^*(x'_n, \Gamma(\omega)) = \delta^*(x', \Gamma(\omega))$$

By condition (ii), there exists a positive integrable function h_A such that

$$\sup_{y' \in A} |\delta^*(y', \Gamma(\omega))| \leq h_A(\omega)$$

Whence

$$\lim_{n \to \infty} \int_\Omega \delta^*(x'_n, \Gamma(\omega))\mu(d\omega) = \int_\Omega \delta^*(x', \Gamma(\omega))\mu(d\omega)$$

by the Lebesgue convergence theorems. That achieves the proof.

16 - Corollary V-16 - Let E be a quasi-complete locally convex Suslin space and f a scalarly integrable mapping from Ω to E. Let U be the unit ball of $L^\infty_{\mathbb{R}}(\Omega, \mathcal{A}, \mu)$. Assume the following conditions: (i) For every g in $L^\infty_{\mathbb{R}}(\Omega, \mathcal{A}, \mu)$ the weak integral $\int gf \, d\mu$ belongs to E (ii) For every fixed equicontinuous subset A in E' there exists a positive integrable function h_A such that

$$\sup_{x' \in A} |\langle x', f(\omega) \rangle| \leq h_A(\omega)$$

<u>for all</u> $\omega \in A$.

<u>Then, we have the following properties</u> :

1) $\{\int g\, f\, d\mu | g \in U\}$ <u>is convex and compact in</u> E.

2) <u>If</u> μ <u>is non atomic</u>, <u>the closure of the set</u> $\{\int_A f\, d\mu | A \in \mathcal{A}\}$ <u>is equal to the convex compact set</u> $\{\int gf\, d\mu | g \in U^+\}$

<u>Proof</u> 1) Let $K = [-1,1] \subset \mathbb{R}$ and put $h(\omega,r) = rf(\omega)$, $r \in \mathbb{R}$, $\omega \in \Omega$. Then the multifunction Γ defined on Ω with non empty convex compact values in E by

$$\Gamma(\omega) = h(\omega, K) \quad , \quad \forall\, \omega \in \Omega$$

is clearly $\mathcal{A}$-measurable and by theorem III-38, a $\mathcal{A}$-measurable selection of Γ is necessarily of the form $\omega \to g(\omega)f(\omega)$ where g is a $\mathcal{A}$-measurable function with values in K. Moreover, it is elementary to check that

$$\sup_{x' \in A} \sup_{r \in K} |< x', rf(\omega) >| \leq h_A(\omega), \ \forall\, \omega \in \Omega$$

Hence, the weak integral

$$\int \Gamma\, d\mu = \{\int h\, d\mu | h \in S_\Gamma\} = \{\int gf\, d\mu | g \in U\}$$

is convex and compact in E by virtue of theorem V-15.

2) If μ is non atomic, then we have already observed (See remark of theorem V-12) that the set $\ddot{U}^+$ of extreme points of U^+, that is ,

$$\ddot{U}^+ = \{\chi_A | A \in \mathcal{A}\}$$

is dense in U for the weak topology $\sigma(L^\infty_{\mathbb{R}}(\Omega,\mathcal{A},\mu), L^1_{\mathbb{R}}(\Omega,\mathcal{A},\mu))$. As the linear mapping $g \to \int gf\, d\mu$ from $L^\infty_{\mathbb{R}}(\Omega,\mathcal{A},\mu)$ to E is continuous with respect to the topologies $\sigma(L^\infty_{\mathbb{R}}, L^1_{\mathbb{R}})$ and $\sigma(E,E')$, we conclude easily that the set $\{\int \chi_A\, f\, d\mu | A \in \mathcal{A}\}$ is dense in the convex compact set $\{\int gf\, d\mu | g \in U^+\}$ with respect to the topology $\sigma(E,E')$. So the closure of the set $\{\int_A f\, d\mu | A \in \mathcal{A}\}$ is equal to the convex compact set $\{\int gf\, d\mu | g \in U^+\}$

Remark. In the situation of corollary V-16, only the compactness property of the set $\{\int gf\, d\mu \mid g \in U\}$ is not obvious. However, in this special case, there are at least two proofs which yield the preceding compactness property when E is a Banach space. Namely, if $f \in L^1_E(\Omega, \mathcal{A}, \mu)$, then the linear operator $T : g \to \int gf\, d\mu$ from $L^\infty_{\mathbb{R}}(\Omega, \mathcal{A}, \mu)$ to E transforms the unit ball U of $L^\infty_{\mathbb{R}}(\Omega, \mathcal{A}, \mu)$ into a convex compact set of E. This fact was stated by Grothendieck ([11], theorem 5, p.155) by using the special properties of $L^1_{\mathbb{R}}(\Omega, \mathcal{A}, \mu)$; on the other hand to prove the compactness of $T(U)$, one may observe that the operator T from the Banach space $L^\infty_{\mathbb{R}}(\Omega, \mathcal{A}, \mu)$ to the Banach space E is nuclear ([17]), that is, T admits a representation of the form

$$T(g) = \sum_{n=1}^{\infty} \lambda_n < \nu_n, g > x_n, \ \forall g \in L^\infty_{\mathbb{R}}(\Omega, \mathcal{A}, \mu) \ \text{ with } \ (\lambda_n) \in \ell^1,$$

$\nu_n \in L^\infty_{\mathbb{R}}(\Omega, \mathcal{A}, \mu)'$, $\|\nu_n\| \leq 1$, $\|x_n\| \leq 1$.

Indeed, it is well known (see, for instance Trèves [17], theorem III. 45.1) that f admits a representation of the form

$$f = \sum_{n=1}^{\infty} \lambda_n \, h_n \otimes x_n$$

with $(\lambda_n) \in \ell^1$, $N_1(h_n) \leq 1$, $\|x_n\| \leq 1$. For every $g \in L^\infty_{\mathbb{R}}(\Omega, \mathcal{A}, \mu)$, put

$$< \nu_n, g > = \int g\, h_n \, d\mu$$

Then, clearly $\nu_n \in L^\infty_{\mathbb{R}}(\Omega, \mathcal{A}, \mu)'$ and $\|\nu_n\| \leq 1$. Since a nuclear mapping is a compact mapping, $T(U)$ is compact.

§ 5 - INTEGRAL REPRESENTATION THEOREM OF MULTIFUNCTIONS FROM A KÖTHE SPACE TO A LOCALLY CONVEX SUSLIN SPACE.

We give here an integral representation theorem for multifunctions which is based on the compactness theorem V-14 and is a multivalued version of the well known Dunford-Pettis's theorem (See for instance

Tulcea's book [18]). Recall that the weak integral of a scalarly integrable multifunction Γ is defined by $\int \Gamma \, d\mu = \{\int f \, d\mu \mid f \in S_\Gamma\}$ (See remark of theorem V-14).

17 - Theorem V-17 - Let $(\Omega, \mathcal{A}, \mu)$ be a complete σ-finite positive measure space, E a locally convex Suslin space. Let M be a multifunction from a Köthe space Λ^* with non empty convex compact values in E satisfying the following conditions :

(i) For every fixed $x' \in E'$ and every fixed $A \in \mathcal{A}$ with $\mu(A) < +\infty$, the scalar function $f \to \delta^*(x', M(f))$ is continuous on the Banach space $L^1(A, \mathcal{A}|_A, \mu|_A)$ where $\mathcal{A}|_A$ is the tribe induced by $\mathcal{A}$ on A and $\mu|_A$ is the restriction of μ to A.

(ii) $M(f+g) = M(f) + M(g)$ if $fg \geq 0$; $f, g \in \Lambda^*$.

(iii) $M(\lambda f) = \lambda \, M(f)$, $\lambda \in \mathbb{R}$, $f \in \Lambda^*$.

(iv) There exists a countable $\mathcal{A}$-measurable partition (A_n) of Ω with $0 < \mu(A_n) < +\infty$ and a sequence of balanced convex compact sets (Q_n) such that

$$\begin{cases} \forall f \in \Lambda^*, \ \forall n \in \mathbb{N}, & M(\chi_{A_n} f) \subset N_1(\chi_{A_n} f) Q_n \\ \forall x' \in E', \ \forall f \in \Lambda^*, & \lim\limits_{m \to \infty} \delta^*(x', M(\sum\limits_{n=1}^{m} \chi_{A_n} f)) = \delta^*(x', M(f)) \end{cases}$$

Then, there exists a scalarly $\mathcal{A}$-measurable multifunction defined on Ω with non empty convex compact values in E such that

(a) $\forall x' \in E'$, $\delta^*(x', \Gamma(.)) \in \Lambda$

(b) $\forall f \in \Lambda^*$, $M(f) = \int f \Gamma \, d\mu = \{\int fg \, d\mu \mid g \in \mathcal{S}_\Gamma\}$ where $\int f\Gamma \, d\mu$ is the weak integral of the scalarly integrable multifunction $f\Gamma$ and $\mathcal{S}_\Gamma$ is the set of all $\mathcal{A}$-measurable selections of Γ.

<u>Proof</u>. For every $\mathcal{A}$-measurable set Z in A_n, let us put

$$L(Z) = M(\chi_Z)$$

Then L is additive by condition (ii). For each n, there exists by condition (iv) a balanced convex compact set Q_n such that, for every $\mathcal{A}$-measurable set $A \subset A_n$, we have

$$\forall x' \in E', \quad \delta^*(x', L(A)) \leq \mu(A)\delta^*(x', Q_n)$$

This implies that each scalar measure $m_{x'} : A \to \delta^*(x', L(A))$ defined on the finite measure space $(A_n, \mathcal{A}_n, \mu|_{A_n})$ (where $\mathcal{A}_n$ is the tribe induced by $\mathcal{A}$ on A_n) is bounded and continuous with respect to the restriction $\mu|_{A_n}$ of μ to A_n.

Therefore, by virtue of the Radon-Nikodym theorem for scalar measure, each m_x admits a density $\dot{g}_{x'} \in L^\infty(A_n, \mathcal{A}_n, \mu|_{A_n})$ such that

$$\begin{cases} \dot{g}_{x'+y'} \leq \dot{g}_{x'} + \dot{g}_{y'} : x' \in E',\ y' \in E' \\ \dot{g}_{\lambda x'} = \lambda\, \dot{g}_{x'}, \quad \lambda \geq 0. \end{cases}$$

Let ρ be a positive linear lift of $L^\infty(A_n, \mathcal{A}_n, \mu|_{A_n})$. Such a lift exists by Ionescu Tulcea's theorems (see for instance Meyer. p. 195 and Tulcea's book). Let $g_{x'} = \rho(\dot{g}_{x'})$.

Then, we have, for every $\omega \in \Omega$,

$$\begin{cases} g_{x'+y'}(\omega) \leq g_{x'}(\omega) + g_{y'}(\omega),\ x' \in E',\ y' \in E' \\ g_{\lambda x'}(\omega) = \lambda\, g_{x'}(\omega), \quad \lambda \geq 0. \end{cases}$$

Let us define the convex compact set $\Gamma_n(\omega)$ by

$$\delta^*(x', \Gamma_n(\omega)) = g_{x'}(\omega), \quad \forall x' \in E', \quad \forall \omega \in A_n$$

Then, we have, for every $\mathcal{A}$-measurable set $Z \subset A_n$, and every $x' \in E'$,

$$\delta^*(x', L(Z)) = \int_Z \delta^*(x', \Gamma_n(\omega))\mu(d\omega).$$

Denote by $\mathcal{S}_{\Gamma_n}$ the set of all $\mathcal{A}_n$-measurable selections of Γ_n.

Let us define $\Gamma(\omega) = \Gamma_n(\omega)$ if $\omega \in A_n$. We first prove that for every

$h \in \Lambda^*$, the weak integral $\int_{A_n} h\Gamma_n \, d\mu$ of the multifunction $\chi_{A_n} h \Gamma_n$

$$M(\chi_{A_n} h) = \int_{A_n} h \Gamma_n \, d\mu = \{\int_{A_n} hg \, d\mu | g \in \mathcal{S}_{\Gamma_n}\}$$

is equal to $M(\chi_{A_n} h)$. Thanks to the theorem V-14, it is equivalent to prove

$$\forall x' \in E', \quad \delta^*(x', M(\chi_{A_n} h)) = \int_{A_n} \delta^*(x', h(\omega)\Gamma_n(\omega))\mu(d\omega)$$

By condition (iv), we have, for every $h \in \Lambda^*$,

$$\delta^*(x', M(\chi_{A_n} h)) \leq N_1(\chi_{A_n} h) \; \delta^*(x', Q_n)$$

By condition (i), the restriction of the functions $f \to \delta^*(x', M(f))$ to $L^1(A_n, \mathcal{Q}_n, \mu|_{A_n})$ (where $\mathcal{Q}_n$ is the tribe induced by $\mathcal{Q}$ on A_n and $\mu|_{A_n}$ is the restriction of μ to A_n) is continuous with respect to the norm topology of $L^1(A_n, \mathcal{Q}_n, \mu|_{A_n})$. Moreover, by condition (ii), for every simple function of the form

$$s = \sum_{i=1}^{k} b_i \chi_{B_i}$$

with B_i $\mathcal{Q}$-measurable and $B_i \subset A_n$, we have

$$M(\chi_{A_n} s) = \int_{A_n} s(\omega) \; \Gamma(\omega) \; \mu(d\omega)$$

Since the set of simple functions is densee in $L^1(A_n, \mathcal{Q}_n, \mu|_{A_n})$ we deduce easily by the continuity of $f \to \delta^*(x', M(f))$ on $L^1(A_n), \mathcal{Q}_n, \mu|_{A_n})$ that, for every $h \in \Lambda^*$, the following equalities hold

$$\delta^*(x', M(\chi_{A_n} h)) = \int_{A_n} \delta^*(x', h(\omega) \; \Gamma(\omega))\mu(d\omega)$$

Now, it is clear that the set $\mathcal{S}_\Gamma$ of all scalarly $\mathcal{Q}$-measurable selections of Γ is not empty. For every fixed $f \in L^\infty(\Omega, \mathcal{Q}, \mu)$, every fixed $h \in \Lambda^*$ and every fixed $g \in \mathcal{S}_\Gamma$, we have

$$\int (\sum_{n=\ell}^{m} \chi_{A_n} \; f \; hg) \, d\mu \in M(\sum_{n=\ell}^{m} \chi_{A_n} \; f \; h)$$

by definition of Γ_n and condition (ii).

This implies,

$$\forall x' \in E', \; \delta^*(-x', M(\sum_{n=\ell}^{m} \chi_{A_n} \; f \; h)) \leq \int < \sum_{n=\ell}^{m} \chi_{A_n} \; hg, \; x' > f \, d\mu$$

$$\leq \delta^*(x', M(\sum_{n=\ell}^{m} \chi_{A_n} \; f \; h))$$

By using condition (i) and (iv), we conclude that, for every fixed $x' \in E'$ and $h \in \Lambda^*$, the sequence $(< \sum_{n=1}^{m} \chi_{A_n} hg, x'>)$ is a $\sigma(L^1, L^\infty)$ Cauchy sequence. But the pointwise limit on Ω of this sequence is $< hg, x' >$, therefore, by a classical property of L^1 space (Dunford-Schwartz, p.295) we have

$$\lim_{m \to \infty} < \sum_{n=1}^{m} \chi_{A_n} hg, x' > = < hg, x' >$$

for the norm topology of $L^1(\Omega, \mathcal{A}, \mu)$. This shows that $S_\Gamma \subset \Lambda_E$ and $\int hg \, d\mu \in M(h)$ for every $g \in S_\Gamma$ and every $h \in \Lambda^*$. Given a fixed element $x' \in E'$, there exists, by the measurable implicit function theorem III-38, an element $\sigma \in \mathcal{S}_\Gamma$ such that

$$< x', \sigma(\omega) > = \delta^*(x', \Gamma(\omega)), \quad \forall \omega \in \Omega$$

This proves that, for every fixed $x' \in E'$, $\delta^*(x', \Gamma(.))$ belongs to Λ. Now, by virtue of theorem V-13, S_Γ is convex and compact for the topology $\sigma(L_{E'}, A^* \otimes E')$; for every fixed element h in Λ^*,

$$\{\int hg \, d\mu \mid g \in S_\Gamma\} = \int h \, \Gamma \, d\mu$$

is a convex, compact set included in the convex compact set M(h).

It remains to prove that

$$M(h) = \{\int hg \, d\mu \mid g \in S_\Gamma\} , \quad \forall h \in \Lambda^*.$$

Let $\xi \in M(h) \setminus \int h \, \Gamma \, d\mu$. By the Hahn-Banach theorem, there exists $x' \in E'$ such that

$$\forall g \in \mathcal{S}_\Gamma, \int_\Omega < x', g(\omega) > h(\omega) \, \mu(d\omega) < < x', \xi > \leq \delta^*(x', M(h))$$

By applying again the above implicit measurable function theorem, there exists an element $\tilde{g} \in \mathcal{S}_\Gamma$ such that

$$< x', \tilde{g}(\omega) > h(\omega) = \delta^*(x', h(\omega) \, \Gamma(\omega)), \quad \forall \omega \in \Omega$$

This implies by integrating on each A_n

$$\int_{A_n} < x', \tilde{g}(\omega) > h(\omega) \, \mu(d\omega) = \int_{A_n} \delta^*(x', h(\omega) \, \Gamma(\omega)) \, \mu(d\omega)$$

$$= \delta^*(x', M(\chi_{A_n} h))$$

By taking account of the condition (iv), we have

$$\delta^*(x', M(h)) = \lim_{m \to \infty} \int_\Omega < x', \sum_{n=1}^{m} \chi_{A_n}(\omega)\, \tilde{g}(\omega) > h(\omega)\, \mu(d\omega)$$

$$= \int_\Omega < x', \tilde{g}(\omega) > h(\omega)\, \mu(d\omega)$$

by virtue of Lebesgue's theorem. This contradicts the inequality

$$\int_\Omega < x', \tilde{g}(\omega) > h(\omega)\, \mu(d\omega) < \delta^*(x', M(h))$$

and completes the proof.

§ 6 - CHARACTERIZATION OF A CLASS OF ABSOLUTELY p SUMMING OPERATORS

In the following we aim to investigate a class of linear operators T from a separable Banach space F to the Banach space $L^p(\Omega, \mathcal{A}, \mu)$ such that

$$|T\, x(\omega)| \leq g(\omega)\, \|x\|$$

for all $\omega \in \Omega$, $x \in F$, where g is a fixed element in $L^p(\Omega, \mathcal{A}, \mu)$ $(1 \leq p < +\infty)$.

18 - Theorem V-18 - Let $(\Omega, \mathcal{A}, \mu)$ be a σ-finite positive measure space, F a separable Banach space, Γ a scalarly $\mathcal{A}$-measurable multifunction from Ω with non empty convex $\sigma(F', F)$ compact values in F' such that the scalar function

$$\omega \to |\Gamma(\omega)| = \sup_{\|x\| \leq 1} |\delta^*(x, \Gamma(\omega))|$$

belongs to $L^p(\Omega, \mathcal{A}, \mu)$ $(1 \leq p < +\infty)$. Denote by S_Γ the quotient of the set $\mathcal{S}_\Gamma$ of all scalarly $\mathcal{A}$-measurable mappings f from Ω to the weak dual F'_s of F' such that $f(\omega) \in \Gamma(\omega)$ a.e and $\mathcal{H}$ the set of all linear mappings A from F to $L^p(\Omega, \mathcal{A}, \mu)$ such that $Ax(.) \leq \delta^*(x, \Gamma(.))$ for all $x \in F$. Then

1) the affine mapping $\sigma \to A$ from S_Γ to $\mathcal{H}$ defined by

$$< \sigma(.), x > = Ax(.)\ , \quad \forall\, x \in F$$

is one-to-one.

2) S_Γ is convex and compact with respect to the topology of convergence on $L^q \otimes F$ $(\frac{1}{p} + \frac{1}{q} = 1)$

3) In addition, the elements A in $\mathcal{H}$ are absolutely p-summing operators in the following sense ([13]) : For any finite sequence $(x_i)_{1 \leq i \leq n}$ in F, we have

$$\sum_{i=1}^{n} [N_p(Ax_i)]^p \leq (\int_\Omega |\Gamma(\omega)|^p \mu(d\omega)) | \sup_{\|x'\| \leq 1} (\sum_{i=1}^{n} |< x', x_i >|^p)|$$

where N_p is the norm of the space $L^p(\Omega, \mathcal{Q}, \mu)$.

Proof. First we remark that F'_s is a Lusin space because the unit ball of F' is a compact metrizable space if we endow this unit ball with the weak topology $\sigma(F',F)$. Moreover the dual of F'_s is clearly F. Second, the pair (L^p, L^q) is a Köthe pair space so that properties 1) and 2) of theorem V-18 are easy consequences of the proof of theorem V-13 by taking $E = F'_s$.

Let us verify property 3). If A is an element of $\mathcal{H}$, then there exists by property 1) an element $\sigma \in S_\Gamma$ such that

$$< \sigma(\cdot), x > = Ax(.), \quad \forall x \in F$$

By hypothesis we have

$$\Gamma(\omega) \subset |\Gamma(\omega)| B', \quad \forall \omega \in \Omega$$

where B' is the unit ball of F'. We deduce easily that

$$|< \sigma(\omega), x >| \leq |\Gamma(\omega)| \, \|x\| \qquad \text{a.e.}$$

because $\delta^*(x,B')$ is equal to $\|x\|$. Now, if $(x_i)_{1 \leq i \leq x}$ is any finite sequence in F, then

$$\sum_{i=1}^{n} |Ax_i(\omega)|^p = \sum_{i=1}^{n} |< \sigma(\omega), x_i >|^p \quad , \quad \forall \omega \in \Omega$$

where σ is a scalarly $\mathcal{Q}$-measurable mapping from Ω to F'_s such that $\sigma(\omega) \in \Gamma(\omega) \subset |\Gamma(\omega)| B'$ a.e. This implies that

$$\sum_{i=1}^{n} |<\sigma(\omega),\ x_i>|^p \le (\sup_{\|x'\|\le 1} \{\sum_{i=1}^{n} |<x',x_i>|^p\})|\Gamma(\omega)|^p$$

for all ω. Integrating, we have

$$\sum_{i=1}^{n} [N_p(Ax_i)]^p \le (\int_\Omega |\Gamma(\omega)|^p \mu(d\omega)) \sup_{\|x'\|\le 1} \{\sum_{i=1}^{n} |<x',x_i>|^p\}$$

Hence, A is absolutely p-summing.

Remark. Theorem V-18 shows that every linear operator T from a separable Banach space F to the Banach space $L^p(\Omega,\mathcal{A},\mu)$ such that $|Tx(\omega)| \le g(\omega)\|x\|$ for all $\omega \in \Omega$, $x \in F$ with $g \in L^p(\Omega,\mathcal{A},\mu)$ is an absolutely p-summing operator. This fact can be obtained directly by using the methods we used in theorem V-13 and theorem V-17. However in theorem V-13, we show more ; namely properties 1) and 2) of theorem V-18 and theorem IV-15 allow us to establish the relations between the sets of extreme points of S_Γ, $\mathcal{H}$ and $\mathcal{H}^*$ where $\mathcal{H}^* = \{A^* : L^q(\Omega,\mathcal{A},\mu) \to F' \mid A \in \mathcal{H}\}$ and A^* is the adjoint of A ; we leave the details of proofs of these facts to the reader.

Note. In certain situations, the set S_Γ of all scalarly integrable selections of a multifunction Γ from Ω with convex closed values in a locally convex space is closed for some adequate topologies. For instance, if Γ is a multifunction from Ω with non empty convex closed values in a Banach space F, then it is easy to prove that the set S_Γ of all integrable selections of Γ is convex and closed in the Banach space $L^1_E(\Omega,\mathcal{A},\mu)$. This is, however, a standard property. Now, we aim to indicate a "closure property" of the set of all scalarly integrable selections of a scalarly $\mathcal{A}$-measurable multifunction with convex closed values in some non standard cases.

§ 7 - CLOSURE THEOREM OF THE SET OF MEASURABLE SELECTIONS OF A MEASURABLE MULTIFUNCTION.

19 - Proposition V-19 - Let F be a separable Banach space, F'_s the weak dual of F and Γ a multifunction from Ω with non empty convex closed weakly locally compact values in F' which contain no line. Let B_Γ (resp. L_Γ) be the quotient of the set $\mathcal{B}_\Gamma$ (resp. $\mathcal{L}_\Gamma$) of all scalarly $\mathcal{Q}$-measurable and bounded (resp. scalarly integrable) selections of Γ. Then, B_Γ (resp. L_Γ) is closed with respect to the topology $\sigma(L^\infty_{F'_s}, L^1_F)$ (resp. the topology of convergence on $L^\infty \otimes F$).

Proof. We prove only the closure property of B_Γ because the proof of this property can be applied to L_Γ. We repeat the construction given in the proof of theorem V-1. Let $(\dot{v}_i, i \in D)$ be a generalized sequence in B_Γ which converges to $\dot{\tilde{v}}$ with respect to the topology $\sigma(L^\infty_{F'_s}, L^1_F)$:

$$\lim_{i \in D} < \dot{v}_i, f > = < \dot{\tilde{v}}, f > , \quad \forall f \in L^1_F$$

Let us prove that $\dot{\tilde{v}} \in B_\Gamma$. Suppose that $\dot{\tilde{v}} \notin B_\Gamma$. Then, by lemma III-32 and lemma III-34, the set A of the points ω such that $\tilde{v}(\omega) \notin \Gamma((\omega)$ is $\mathcal{Q}$-measurable and there exists a sequence $(e_n)_{n\in\mathbb{N}}$ in F such that

$$A = \bigcup_{n\in N} \{\omega \in \Omega | \delta^*(e_n, \Gamma(\omega)) < < e_n, \tilde{v}(\omega)>\}$$

So there exists $e_n \in E$ such that the set

$$A_n = \{\omega \in \Omega | \delta^*(e_n,\Gamma(\omega)) < < e_n, \tilde{v}(\omega) >\}$$

satisfies $0 < \mu(A_n)$. Integrating on A_n, we have

$$\int_{A_n} < e_n, \tilde{v} > d\mu = \lim_{i \in D} \int_{A_n} < e_n, v_i > \leq \int_{A_n} \delta^*(e_n,\Gamma(\omega))\mu(d\omega)$$

$$\int_{A_n} \delta^*(e_n, \Gamma(\omega))\mu(d\omega) < \int_{A_n} < e_n,\tilde{v} > d\mu$$

So the first inequality contradicts the second inequality which completes

BIBLIOGRAPHY OF CHAPTER V

1 - BOURBAKI, N. : Integration vectorielle. chapitre 6. Hermann Paris 1959

2 - BISMUT, J.M. : Intégrales convexes et Probabilités. Journal Math. Analysis and Applications 42, 639-673 (1973).

3 - CASTAING, Ch : Quelques applications du théorème de Banach Dieudonné à l'Intégration. Faculté des Sciences Montpellier Publication n°67 (1969-1970).

4 - CASTAING, Ch. : Quelques résultats de compacité liés à l'Intégration. C.R. Acad. Sc. Paris 270, 1732-1735 (1970) et Bulletin Soc. Math. France, 31, 73-81 (1972).

5 - CASTAING, Ch. : Le théorème de Dunford-Pettis généralisé. Faculté des Sciences Montpellier. Publication n°43 (1968-1969) et C.R. Acad. Sc. Paris 268, 327-329 (1969).

6 - CASTAING, Ch. : Proximité et mesurabilité. Un théorème de compacité faible. Colloque sur la théorie mathématique du contrôle optimal, Bruxelles, 1969.

7 - CASTAING, Ch., VALADIER, M. : Equations différentielles multivoques dans les espaces vectoriels localement convexes. Revue Informatique et de Recherche Opérationnelle 16, 3-16 (1969).

8 - CLAUZURE, P. : Dualité et compacité dans les espaces de Köthe généralisés. C.R. Acad. Sc. Paris 278, 1710-1713 (1972).

9 - DEBREU, G., SCHMEIDLER, D. : The Ralon Nikolym derivative of a correspondence. Proc. Sixth Berkeley Symposium on mathematical Statistics and Probability 1971.

10 - GROTHENDIECK, A. : Espaces vectoriels topologiques. Sao Paulo.

11 - GROTHENDIECK, A. : Sur les applications linéaires faiblement compactes du type C(K). Canadian Journal of Math. 5, 129-173 (1953).

12 - MEYER, P.A. : Probabilités et Potentiel. Hermann Paris 1966.

13 - PIETSCH, A. : Nuckleare lokalconvexe Raüme, Berlin : Akad-Verl. 1965.

14 - ROCKAFELLAR, R.T. : Integrals which an convex functionals II. Pacific Journal of Math. 39-2, 439-469 (1971).

15 - ROCKAFELLAR, R.T : Convex integrals functionals and duality. Contributions to non linear functional analysis. Academic press, 215-236 (1971).

16 - STRASSEN, V. : The existence of probability measures with given marginals. Ann. Math. Stat. 38, 423-439 (1965).

17 - TREVES, F : Topological vector spaces, distributions, and kernels. Academic Press 1967.

18 - TULCEA, A., TULCEA. C. : Topics in the theory of liftings. Springer-Verlag. Berlin Heidelberg New York 1969.

19 - VALADIER, M. : Un théorème d'inf-compacité. Exposé n°14, Séminaire d'Analyse Convexe Montpellier 1971.

20 - VALADIER, M. : Sur le théorème de Strassen. C.R. Acad. Sc. Paris, 278, (1974)et Exposé n°4, Séminaire d'Analyse convexe Montpellier 1974.

CHAPTER VI

PRIMITIVE OF MULTIFUNCTIONS AND MULTIVALUED DIFFERENTIAL EQUATIONS.

The theory of multivalued differential equations comes from the theory of optimal control. The important point of the study of multivalued differential equations is the compactness of trajectories from which we can derive the existence of optimal controls. For this purpose we begin to give below the notion of a primitive of a multifunction and its properties which will be used later and are of independent interest.

§ 1 - PRIMITIVE OF MULTIFUNCTIONS

Let E be a real Hausdorff locally convex space and $\mathcal{C}_E[0,T]$ the vector space of all continuous mappings from the interval $[0,T]$ of $\mathbb{R}$ into E. We shall endow $\mathcal{C}_E[0,T]$ with the topology of uniform convergence on $[0,T]$. Let ds be the Lebesgue measure on $[0,T]$ and Γ a scalarly ds-integrable mapping from $[0,T]$ with non empty convex compact values in E (i.e. for every x' in E', the scalar function $\delta^*(x',\Gamma(.))$ is ds-integrable). Denote by $\mathcal{S}_\Gamma$ the set of scalarly ds-integrable selections of Γ and define the weak integral $\int_t^{t'} \Gamma(s)ds$ of Γ on $[t,t']$ by

$$\int_t^{t'} \Gamma(s)ds = \{\int_t^{t'} f(s)ds \mid f \in \mathcal{S}_\Gamma\}$$

where $\int_t^{t'} f(s)ds$ is the weak integral of the scalarly ds-integrable function f on $[t,t']$.

Denote by E_σ the vector space E endowed with the weak topology $\sigma(E,E')$.

1 - Theorem V-1. Suppose that E is Suslin and let Γ be a scalarly ds-integrable multifunction from $[0,T]$ with non empty convex compact values in E such

<u>that for every scalarly integrable selection</u> f <u>of</u> Γ, <u>the weak integral</u> $\int_0^t f(s)\,ds$ $(t \in [0,T])$ <u>belongs to</u> E. <u>Let</u> M <u>be a compact subset of</u> E. <u>Denote by</u>

$$\mathcal{X} = \{X : [0,T] \to E \mid X(0) \in M,\ X(t) = X(0) + \int_0^t X'(s)ds,\ X' \in \mathcal{S}_\Gamma\}$$

<u>Then,</u> $\mathcal{X}$ <u>is a compact subset of</u> $\mathcal{C}_{E_\sigma}[0,T]$.

<u>Proof</u>. We have to show first that $\mathcal{X}$ is relatively compact in $\mathcal{C}_{E_\sigma}[0,T]$. By virtue of Ascoli's theorem, it suffices to proves that $\mathcal{X}$ is an equicontinuous subset of $\mathcal{C}_{E_\sigma}[0,T]$ and for every fixed $t \in [0,T]$, the set $\mathcal{X}(t) = \{X(t) \mid X \in \mathcal{X}\}$ is relatively $\sigma(E, E')$ compact. Let us verify that $\mathcal{X}$ is equicontinuous. For $0 \le t \le t' \le T$, we have

$$\forall X \in \mathcal{X},\ X(t') - X(t) \in \int_t^{t'} \Gamma(s)\,ds = \{\int_t^{t'} f(s)\,ds \mid f \in \mathcal{S}_\Gamma\}$$

By theorem V-14, we can write

$$\forall X \in \mathcal{X},\ \forall e' \in E',\ \langle e', X(t') - X(t) \rangle \le \int_t^{t'} \delta^*(e', \Gamma(s))ds$$
$$= \delta^*(e', \int_t^{t'} \Gamma(s)ds)$$

It follows immediately from theorem II-20 that the multifunction $t' \to \int_t^{t'} \Gamma(s)ds$ is upper semi-continuous for the weak topology $\sigma(E,E')$ when t is fixed. Now, given a weak neighbourhood V of origin in E and a number $t \in [0,T]$, there exists, by using the upper-semi-continuity property of the multifunction $t' \to \int_t^{t'} \Gamma(s)\,ds$, a positive number ε such that

$$t' \in [t, t+\varepsilon] \Rightarrow \int_t^{t'} \Gamma(s)\,ds \subset V.$$

This proves that $\mathcal{X}$ is "equicontinuous on the right of t" and the above arguments proves that $\mathcal{X}$ is also "equicontinuous on the left of t" The fact that $\mathcal{X}$ is relatively $\sigma(E,E')$ compact is trivial since for every $t \in [0,T]$, and every $X \in \mathcal{X}$,

$$X(t) \in M + \int_0^t \Gamma(s)\,ds$$

and the last set is $\sigma(E,E')$ compact by theorem V-14. Now, let $\mathcal{U}$ be an ultrafilter in $\mathfrak{X}$ converging to an element $\tilde{X} \in \mathcal{C}_{E_\sigma}[0,T]$. For every $t \in [0,T]$ and every $e' \in E'$, we have (since $M \times S_\Gamma$ and $\mathfrak{X}$ are isomorphic)

$$< e', \tilde{X}(t) - \tilde{X}(0) > = \lim_{\mathcal{U}} < e', X(t) - X(0) >$$

$$= \lim_{\mathcal{U}} < e', \int_0^t X'(s)ds >, \; X' \in \mathcal{S}_\Gamma$$

By theorem V-14, there exists $f \in \mathcal{S}_\Gamma$ such that

$$\forall\, e' \in E', \; \forall\, h \in \mathcal{L}^\infty([0,T],ds), \; \lim_{\mathcal{U}} < X', h \otimes e' > = < f, h \otimes e' >$$

Hence it follows that

$$\forall\, t \in [0,T], \; \forall\, e' \in E', \; \lim < e', \int_0^t X'(s)ds > = < e', \int_0^t f(s)ds >$$

Therefore we have

$$\tilde{X}(t) - \tilde{X}(0) = \int_0^t f(s)ds$$

This completes the proof.

2 - The notion of a primitive of a scalarly integrable multifunction is first introduced by Ghouila-Houri in the case when E has finite dimension and this notion is illustrated by the following theorem which generalizes the property of the primitive of an integrable function.

<u>Definition</u>. Under assumptions and notation of theorem VI-1, we define the <u>primitive</u> $\mathcal{P}_\Gamma$ of Γ by

$$\mathcal{P}_\Gamma = \{X : [0,T] \to E \,|\, X(t'') - X(t') \in \int_{t'}^{t''} \Gamma(s)ds, \; 0 \le t' \le t'' \le T\}$$

2 - <u>Theorem VI-2</u> - <u>With assumptions and notation of theorem</u> VI-1, <u>let</u> $\mathcal{P}_\Gamma$ <u>be the primitive of</u> Γ. <u>Then we have the following equality</u>,

$$\mathfrak{X} = \{X \in \mathcal{P}_\Gamma \,|\, X(0) \in M\}$$

<u>Proof</u>. Let us put $\mathfrak{X}_1 = \{X \in \mathcal{P}_\Gamma \,|\, X(0) \in M\}$. It is obvious that $\mathfrak{X}$ is included in $\mathfrak{X}_1$.

By virtue of theorem VI-1 it suffices to prove that $\mathfrak{X}$ is a dense subset of $\mathfrak{X}_1$ for the topology of pointwise convergence on $[0,T]$. Let X be an element of $\mathfrak{X}_1$ and let $0 = t_0 < t_1 < \ldots < t_n \leq T$ be a subdivision of $[0,T]$. By theorem V-14 there exists $f \in S_\Gamma$ such that

$$\int_{t_i}^{t_{i+1}} f(s)ds = X(t_{i+1}) - X(t_i)$$

Let us define

$$Y(t) = X(0) + \int_0^t f(s)ds, \quad 0 \leq t \leq T.$$

Then

$$Y(t_i) = X(t_i), \quad 0 \leq i \leq n$$

and Y belongs to $\mathfrak{X}$. This achieves the proof.

<u>Remark</u>. If f is a scalarly ds-measurable mapping from $[0,T]$ into the weak dual E'_s of a separable Banach space E and if $|f| : t \to \|f(t)\|$ is ds-integrable, then

$$t \to X(t) = X(0) + \int_0^t f(s)ds \quad (t \in [0,T])$$

is absolutely continuous (with respect to the strong topology placed on E'). Conversely if X is a mapping from $[0,T]$ into E' such that

$$\|X(t) - X(\tau)\| \leq \int_\tau^t g(s)ds, \quad 0 \leq \tau \leq t \leq T.$$

where g is a positive ds-integrable function, then, there exists a scalarly ds-measurable mapping f from $[0,T]$ into E'_s with $|f| \leq g$, ds-integrable, such that

$$X(t) = X(0) + \int_0^t f(s)ds$$

Indeed, this property is an immediate corollary of theorem VI-2 by taking $\Gamma(s) = g(s)B'$ $(s \in [0,T])$ where B' is the unit ball of E'. In other words, X is the primitive of f; moreover X is weakly derivable a.e. and the weak derivative of X is equal to f. In the particular case where E' is strongly separable, X is strongly derivable a.e. and the strong derivative of X is equal to f by virtue of Dunford-Schwartz's result

([10] (theor. 12.8, p.217) (See also Yosida [15]).

§2 - DERIVATION OF MULTIFUNCTION OF "BOUNDED VARIATION".

3 - We now state a multivalued version of the above property.

Theorem VI-3 - *Let* F *be a multifunction from* $[0,T]$ *with non empty convex compact values in the weak dual* E'_s *of a separable Banach space* E, *let* v *be an increasing and left continuous real function defined on* $[0,T]$ *such that*

(i) $|\delta^*(x, F(t)) - \delta^*(x, F(\tau))| \leq \|x\| \; (v(t)-v(\tau)), \; 0 \leq \tau \leq t \leq T, \; x \in E$

(ii) *for every fixed* τ *and every fixed* t *with* $\tau \leq t$, *the function*

$$x \to \delta^*(x,F(t)) - \delta^*(x, F(\tau))$$

is convex, positively homogeneous on E. *Then, there exists a scalarly ds-integrable multifunction* Γ *from* $[0,T]$ *with non empty convex compact values in* E'_s *such that*

(a) $\frac{d}{ds} [\delta^*(x, F(s))] = \delta^*(x, \Gamma(x))$ a.e.

(b) $\delta^*(x,\Gamma(x)) \leq \|x\|\frac{dv}{ds}(s)$ a.e.

where $\frac{dv}{ds}$ *is the derivative of* v.

Proof. For every fixed $x \in E$, condition (i) implies that the real function

$$F_x : t \to \delta^*(x, F(t)) \quad (t \in [0,T])$$

is of bounded variation and left continuous since v is increasing and left continuous. By a well known result concerning bounded variation functions (see also the Gelfand-Dinculeanu theorem for vector bounded variation functions ([9], theorem 1, p.358-383)) there exists a positive Radon measure μ and (for every fixed $x \in E$) a Radon measure m_x such that, for $\tau < t$,

$$\begin{cases} v(t) - v(\tau) = \mu[\tau,t[\\ F_x(t) - F_x(\tau) = m_x[\tau,t[\end{cases}$$

We show now that for every fixed $f \in L^1_+[0,T],\mu)$, the real function $x \to m_x(f)$ is convex and positively homogeneous. Indeed, let $\mathcal{E}$ be the set of positive functions g of the form

$$g = \sum_{i=1}^{m} r_i \mathcal{X}_{I_i}$$

with $r_i \in \mathbb{R}^+$, $I_i = [\tau_i, t_i[$, $I_i \cap I_j = \emptyset$, $i \neq j$. Then, $x \to m_x(g)$ is convex positively homogeneous by condition (ii). Since $\mathcal{E}$ is dense in $L^1_+([0,T],\mu)$, we deduce easily that $x \to m_x(f)$ is convex positively homogeneous; in particular this shows that the "multimeasure" $A \to M(A)$ defined on the Borel tribe $\mathcal{A}$ of $[0,T]$ by formula

$$\delta^*(x, M(A)) = m_x(A), \; x \in E, \; A \in \mathcal{A}$$

is "dominated by μ" ; i.e.,

$$\delta^*(x, M(A)) \leq \mu(A) \; \|x\| \; , \; \forall A \in \mathcal{A} \; , \; \forall x \in E$$

Hence, the Radon measure

$$|M| = \sup_{\|x\| \leq 1} |m_x|$$

is also dominated by μ. Now, by the Lebesgue decomposition theorem, there exist two Borel sets, A and B, with $A \cap B = \emptyset$, $A \cup B = [0,T]$ such that the measure $\mathcal{X}_A|M|$ (resp. $\mathcal{X}_B|M|$) is continuous (resp. singular) with respect to the Lebesgue measure ds. Let us define

$$N = \mathcal{X}_A \, M$$

$$P = \mathcal{X}_B \, M$$

so that $|N| = \mathcal{X}_A|M|$ and $|P| = \mathcal{X}_B|M|$.

Then, for every fixed $x \in E$, the measure $p_x(.) = \delta^*(x, P(.))$ is singular with respect to the Lebesgue measure ds. Hence by virtue of (theor. 12.6, p.214, [10]), the derivative $\frac{dp_x}{ds}(.)$ of the measure p_x with respect to the Lebesgue measure is equal to zero a.e. . Since, the measure

$$|N| = \mathcal{X}_A|M|$$

is ds-continuous and for every ds-measurable set C of $[0,T]$ we have

$$N(C) \subset |N|\,(C)\ B'$$

there exists, by the first part of the proof of theorem V-17 a scalarly $|N|$-measurable multifunction Φ from $[0,T]$ with $\sigma(E',E)$ convex compact non empty values in B' such that, for every ds-measurable set C in $[0,T]$,

$$N(C) = \int_C \Phi(s)d|N|\,(s) = \int_C h(s)\ \Phi(s)\ ds$$

where $|N| = h\ ds$. Then, the multifunction $s \to \Gamma(s) = h(s)\Phi(s)$ is scalarly ds-measurable and satisfies conditions (a) and (b). Indeed, we have, for every $x \in E$, and for every ds-measurable set C,

$$\delta^*(x,\ N(C)) = \int_C \delta^*(x,\ \Gamma(s))\ ds$$

so that the derivative of the measure $n_x(.) = \delta^*(x,\ N(.))$ with respect to the Lebesgue measure ds, is a.e. equal to $\delta^*(x,\ \Gamma(.))$ on $]0,T[$.

But we have, for every $x \in E$ and $0 \le \tau < t \le T$

$$\delta^*(x,\ F(t)) - \delta^*(x,\ F(\tau)) = \delta^*(x,\ M[\tau,t[),$$

so that, the function $\delta^*(x,\ F(.))$ is derivable at $t \in]0,T[$ if and only if the measure $\delta^*(x,\ M(.) = m_x(.)$ is derivable with respect to the Lebesgue measure at t and,

$$\frac{dm_x}{ds}(s) = \frac{d}{ds}\ \delta^*(x,\ F(s))\quad \text{a.e.}$$

Since

$$\frac{dm_x}{ds} = \frac{dn_x}{ds} + \frac{dp_x}{ds} = \frac{dn_x}{ds}\quad \text{a.e.}$$

then

$$\frac{dn_x}{ds}(s) = \frac{d}{ds}\ (\delta^*(x,\ F(s)))\ \text{a.e.}$$

This proves condition (a). Now it suffices to apply condition (i) and the inequality $m_x([\tau,T[) \le \mu([\tau,t[)\|x\|$ to obtain

$$\frac{dm_x}{ds}(s) = \frac{d}{ds}(\delta^*(x, F(s))) \leq \frac{d\mu}{ds}(s)\|x\| = \frac{dv}{ds}(s)\|x\|, \text{ a.e.}$$

Remark. Theorem VI-3 generalizes some results due to Komura ([12]) and Artstein ([1]). The above proof is given by Castaing in ([5]).

§ 3 - CLOSURE THEOREM INVOLVING THE COMPACTNESS PROPERTY OF TRAJECTORIES OF MULTIVALUED DIFFERENTIAL EQUATIONS.

4 - The following theorem will be used to prove the compactness of trajectories of a multivalued differential equation.

Theorem VI-4 - Let U be a topological space and let Φ be a multifunction from $[0,T] \times U$ with non empty convex compact values in a Hausdorff locally convex space E such that for every $t \in [0,T]$, $\Phi(t,.)$ is upper semicontinuous. Let X_n and X belongito $U^{[0,T]}$ and Y_n and Y be scalarly ds-integrable mappings from $[0,T]$ to E. We assume the following hypotheses :

a) There exists a sequence (e'_n) in E' which separates the points of E

b) $\lim\limits_{n \to \infty} X_n(t) = X(t)$ a.e.

c) For every fixed $x' \in E'$, the sequence $(< x', Y_n(.) >)$ converges to $< x', Y(.) >$ with respect to the weak topology $\sigma(L^1[0,T], L^\infty[0,T])$.

d) $Y_n(t) \in \Phi(t, X_n(t))$ a.e.

Then $Y(t) \in \Phi(t, X(t))$ a.e.

Proof. Since $(< x', Y_n(.) >)$ converges weakly to $< x', Y(.) >$, there exists a sequence (Z_m) (where Z_m is a convex combination of $<x', Y_m(.)>$, $< x', Y_{m+1}(.) >, \ldots$) which converges to $< x', Y(.) >$ in the norm topology of $L^1([0,T])$. We can extract from the sequence (Z_m) a subsequence which converges a.e. to $< x', Y(.) >$. By condition b) there exists a null set N such that $X_n(t)$ converges to $X(t)$ for $t \notin N$. Since $\Phi(t,x)$ is convex compact and $\Phi(t,.)$ is upper semi continuous, there exists a null set $N_{x'}$ with $N_{x'} \supset N$ such that

$$< x', Y(t) > \in \bigcap_k \overline{co} (\bigcup_{m \geq k} < x', Y_m(t) >) \subset x' \circ \Phi(t, X(t))$$

for $t \notin N_{x'}$. By lemma III-32, there exists a countable dense subset $\mathcal{F}$ for Mackey topology $\tau(E',E)$ such that the following equivalence holds :

$$< x', Y(t) > \in x' \circ \Phi(t, X(t)), \ \forall\, t \notin \bigcup_{x' \in \mathcal{F}} N_{x'}$$

$$<=>$$

$$Y(t) \in \Phi(t, X(t)) \quad \text{a.e.}$$

This completes the proof.

Remark. The arguments in the following results give us some measurable selections results which will be used in the proof of the existence of solutions of a multivalued differential euqation.

5 - Corollary VI-5 - Suppose that U is a Suslin space and E is a separable Banach space. Let g be a positive ds-integrable function and let K be a balanced convex $\sigma(E,E')$ compact set of E. Let Φ be a multi-function from $[0,T] \times U$ with non empty convex $\sigma(E,E')$ compact values in E such that :

(i) For every $t \in [0,T]$ and every $x \in U$, $\Phi(t,x) \subset g(t) K$

(ii) For every fixed $t \in [0,T]$ and every fixed $x' \in E'$, the scalar function $u \to \delta^*(x',\Phi(t,u))$ is upper semi-continuous on U,

(iii) For every fixed $u \in U$, every fixed $x' \in E'$, the scalar function

$$t \to \delta^*(x', \Phi(t,u))$$

is ds-measurable on $[0,T]$.

Then, for every ds-measurable mapping X from $[0,T]$ into U (i.e. for every Borel set B of U, $X^{-1}(B)$ is ds-measurable), there _exists a scalarly_ ds-_measurable mapping_ Y _from_ $[0,T]$ _into_ E _such that_

$$Y(t) \in \Phi(t, X(t)) \quad \text{a.e.}$$

Proof. Since U is a Suslin space, there exists by theorem III-22 a sequence (X_n) of simple functions from $[0,T]$ into U with countable values such that

$$\forall\, t \in [0,T], \quad \lim_{n \to \infty} X_n(t) = X(t)$$

So, by hypothesis (iii), each multifunction $t \to \Phi(t, X_n(t))$ is scalarly ds-measurable, hence, there exists (theor. III-37), for each n, a scalarly ds-measurable selection Y_n of $t \to \Phi(t, X_n(t))$.
By virtue of (i) and corollary V-4 the sequence (Y_n) is relatively $\sigma(L^1_E\,[0,T],\ L^\infty_{E'_s}\,[0,T])$ compact. Therefore we can extract from (Y_n) a subsequence which converges for the weak topology $\sigma(L^1_E[0,T],\ L^\infty_{E'_s}\,[0,T])$ to an element Y (with $Y(t) \in g(t)\,K$ a.e.). Then, to simplify the notation we can suppose that

$$\forall\, h \in L^\infty_{E'_s}[0,T], \quad \lim_{n \to \infty} < Y_n,\ h > \, = \, < Y,\ h >$$

This implies that, for every $x' \in E'$, the sequence $(< x', Y_n(.) >)$ converges to $< x', Y(.) >$ for the weak topology $\sigma(L^1[0,T],\ L^\infty[0,T])$ so that we can apply theorem VI-4 (since $\Phi(t,.)$ is upper-semi-continuous with respect to the weak topology $\sigma(E,E')$ by virtue of theorem II-20) which implies

$$Y(t) \in \Phi(t, X(t)) \quad \text{a.e.}$$

Remark. There are many variants of corollary VI-5 ([4], [7], [8])
We give here only a more general result which is derived from the properties of measurable multifunctions with convex compact values.

6 - Theorem VI-6 - Let U be a Suslin space and let E be a locally convex Suslin space. Let Φ be a multifunction from $[0,T] \times U$ with non empty convex $\sigma(E,E')$ compact values in E such that

(i) For every fixed $t \in [0,T]$ and every fixed $x' \in E'$, the scalar function $u \to \delta^*(x', \Phi(t,u))$ is upper semi-continuous on U

(ii) For every fixed $u \in U$, every fixed $x' \in E'$, the scalar function $t \to \delta^*(x', \Phi(t,u))$ is ds-measurable in $[0,T]$.

Then, for every ds-measurable mapping X from $[0,T]$ into U (i.e. for every Borel set B of U, $X^{-1}(B)$ is ds-measurable), there exists a scalarly ds-measurable mapping Y from $[0,T]$ into E such that

$$Y(t) \in \Phi(t, X(t)) \quad \text{a.e.}$$

Proof. Since U is a Suslin space, there exists a sequence (X_n) of simple functions from $[0,T]$ into U with countable values such that

$$\forall\, t \in [0,T], \quad \lim_{n \to \infty} X_n(t) = X(t)$$

So, by hypothesis (ii), each multifunction $t \to \Phi(t, X_n(t))$ is scalarly ds-measurable.

Observe that

$$\delta^*(x', \text{co} \bigcup_{n\geq k} \Phi(t,X_n(t))) = \sup_{n\geq k} \delta^*(x', \Phi(t, X_n(t)))$$

and by (i)

$$\emptyset \neq \bigcap_k \overline{\text{co}(\bigcup_{n\geq k} \Phi(t, X_n(t)))} \subset \Phi(t, X(t))$$

By theorem III-37, the multifunction

$$t \to \bigcap_k \overline{\text{co}(\bigcup_{n\geq k} \Phi(t, X_n)t))}$$

is ds-measurable. Hence the multifunction $t \to \bigcap_k \overline{\text{co}(\bigcup_{n\geq k} \Phi(t, X_n(t)))}$ admits by theorem III-22 a scalarly ds-measurable mapping. This finishes our proof.

§ 4 - EXISTENCE THEOREM OF MULTIVALUED DIFFERENTIAL EQUATIONS

7 - We can state now the existence theorem for multivalued differential equations $\dot{X}(t) \in F(t, X(t))$ where F is a multifunction with convex compact values such that $F(t,.)$ is upper-semi-continuous and $F(.,x)$ is ds-measurable and F satisfies an integrability condition generalizing Caratheodory's condition.

Theorem VI-7 - Let E be a locally convex Suslin space, let U be an open set of E_σ and M be a convex compact subset included in U. Let Γ be a scalarly ds-integrable multifunction from $[0,T]$ with non empty convex compact values in E_σ and F be a multifunction from $[0,T] \times U$ with non empty convex compact values in E_σ such that

(i) for every $t \in [0,T]$, every $x \in U$, $F(t,x) \subset \Gamma(t)$

(ii) for every $x \in E$, every $x' \in E'$, $\delta^*(x', F(.,x))$ is ds-measurable

(iii) for every $t \in [0,T]$, every $x' \in E'$, $\delta^*(x', F(t,.))$ is upper-semi-continuous on U.

If, for every t and t' in $[0,T]$, the weak integral $\int_t^{t'} \Gamma(s)ds = \{\int_t^{t'} f(s)ds \mid f \in \mathcal{S}_\Gamma\}$ is included in E, then

a) there exists $T_0 \in]0,T]$ such that $M + \int_0^{T_0} \Gamma(s)\, ds \subset U$

b) for every $\xi \in M$, the set S_ξ of all solutions X defined on $[0,T_0]$ verifying $X(0) = \xi$ is a non empty compact set of $\mathcal{C}_{E_\sigma}[0,T_0]$

c) the multifunction $\xi \to S_\xi$ is upper-semi-continuous from M to $\mathcal{C}_{E_\sigma}[0,T_0]$.

Proof. a) Let V be a neighbourhood of an origin for the weak topology $\sigma(E,E')$ such that $M + V \subset U$. Since we have already proved that the multifunction

$$t \to \int_0^t \Gamma(s)\, ds \quad (t \in [0,T])$$

is upper-semi-continuous from $[0,T]$ into E_σ by virtue of theorem II-20, if V is a neighbourhood of origin in $E\sigma$ such that $M + V \subset U$, , there exists T_0 such that $\int_0^{T_0} \Gamma(s)\, ds \subset V$. So we have $M + \int_0^{T_0} \Gamma(s) ds \subset U$. This proves assertion a).

Now, consider

$$\mathfrak{X} = \{X : [0,T_0] \to E | X(0) \in M,\ X(t) - X(0) = \int_0^t X'(s)ds,\ X' \in S_\Gamma\}$$

By theorem VI-1, $\mathfrak{X}$ is a compact subset of the space $\mathcal{C}_{E_\sigma}[0,T_0]$ and is metrizable since the topology of $\mathfrak{X}$ is equal to the topology of pointwise convergence on a countable dense subset of $[0,T_0]$. For every $X \in \mathfrak{X}$ and every $\xi \in M$, let

$$\Phi(X) = \{Y \in \mathfrak{X} | X(0) = Y(0) = \xi,\ Y'(t) \in F(t,X(t)) \text{a.e.}\}$$

By theorem VI-6, there exists a scalarly ds-measurable mapping σ from $[0,T_0]$ into E such that

$$\sigma(t) \in F(t, X(t)) \text{ a.e.}$$

Since $F(t,X(t)) \subset \Gamma(t)$, σ is scalarly ds-integrable and the mapping

$$t \to Y(t) = X(0) + \int_0^t \sigma(s)\, ds,\ t \in [0,T]$$

belongs to $\mathfrak{X}$ so that $\Phi(X)$ is non empty. We have to show now that the multifunction Φ defined on $\mathfrak{X}$ is upper-semi-continuous in order to apply the Kakutami-Ky-Fan fixed point theorem (see for instance Berge [2]).

Let G be a subset of $M \times \mathfrak{X}^2$ such that $X(0) = Y(0) = \xi$ and $Y'(t) \in F(t, X(t))$ a.e. . It suffices to prove that G is a closed subset of $M \times \mathfrak{X}^2$. Let (ξ_k, X_k, Y_k) be a sequence which converges to $(\xi,X,Y) \in M \times \mathfrak{X}^2$

Then $\quad X(0) = Y(0) = \xi.$

For every k we have

$$Y_k'(t) \in F(t, X_k(t)) \quad \text{a.e.}$$

Since the quotient S_Γ of $\mathcal{S}_\Gamma$ for the equivalence "equality a.e." is compact for the topology of convergence on $L^\infty[0,T_o] \otimes E'$ (cf theorem V-13), S_Γ is a fortiori compact (metrizable!) for the topology of convergence on the space of measurable simple functions $\mathcal{E}_{E'}$ taking a finite number of values in E' (If D is a countable dense subset in $[0,T_o]$ and (e'_n) is a $\sigma(E',E)$ countable dense subset of E', this topology is the coarsest locally convex topology on S_Γ for which the mappings

$$f \to \int_0^t < e'_n, f(s) > ds \quad (n \in N,\ t \in D)$$

are continuous). Therefore we can apply theorem VI-4 to obtain

$$Y'(t) \in F(t, X(t)) \text{ a.e.}$$

This proves that G is closed. Since $\Phi(X)$ is convex and non empty, Φ is upper-semi-continuous and admits a fixed point which is clearly a solution of $X'(t) \in F(t, X(t))$.

c) The last part of the theorem is an easy consequence of the fact thatis G is closed.

8 - We shall now give a global existence theorem for multivalued differential equations of the case where E is a Banach space by using the same methods as in theorem VI-7.

Theorem VI-8 - Let E be the strong dual G'_b of a separable Banach space G. Let F be a multifunction from $[0,T] \times G'$ with non empty convex $\sigma(G',G)$ compact values in G' such that

(i) for every $t \in [0,T]$, $F(t,.)$ is upper-semi-continuous for the weak topology $\sigma(G',G)$

(ii) for every $x \in G'_b$, $F(.,x)$ is scalarly measurable on $[0,T]$, i.e. for every $y \in G$,

$$t \to \delta^*(y, F(t,x))$$

is measurable.

Suppose there exists a positive function h defined on $[0,T] \times \mathbb{R}^+$ satisfying the following conditions

(j) for every $t \in [0,T]$, $h(t,.)$ is increasing and continuous on $\mathbb{R}^+$,

(jj) for every $r \in \mathbb{R}^+$, $h(.,r)$ is integrable on $[0,T]$,

(jjj) for every $t \in [0,T]$, $x \in G'_b$,

$$F(t,x) \subset h(t, \|x\|)B'$$

where B' is the unit ball of G'.

Let r_0 be a strictly positive number. Suppose there exists an absolutely continuous function $\rho : [0,T] \to \mathbb{R}^+$ such that

$$\begin{cases} \rho(t) = r_0 + \int_0^t \rho'(s)ds, & \forall t \in [0,T] \\ \rho'(t) \geq h(t, \rho(t)), & \forall t \in [0,T] \end{cases}$$

Then, for every ξ with $\|\xi\| \leq r_0$, the set S_ξ of all solutions of $\dot{X}(t) \in F(t, X(t))$ a.e. with $X(0) = \xi$ is a non empty compact set in $\mathscr{C}_{G'_s}[0,T]$ and the multifunction $\xi \to S_\xi$ is upper-semi-continuous on the ball $\|\xi\| \leq r_0$ endowed with the weak topology $\sigma(G',G)$.

Proof. Let

$$\Gamma(t) = \rho'(t)B'$$

$$\mathcal{X} = \{X : [0,T] \to G'_s \mid \|X(0)\| \leq r_0,\ X(t) = X(0) + \int_0^t X'(s)ds,\ X' \in S_\Gamma\}$$

and for every $X \in \mathcal{X}$, let us put

$$\Phi(X) = \{Y : [0,T] \to G'_s \mid \|Y(0)\| \leq r_0,\ Y(t) = Y(0) + \int_0^t Y'(s)ds,$$
$$Y'(s) \in F(s, X(s)) \text{a.e.}\}$$

Since each $X \in \mathcal{X}$ is scalarly continuous on $[0,T]$ with values in G'_s, hence, X is a fortiori scalarly measurable, so, we can apply theorem VI-6 because G'_s is Lusin space. So there exists a scalarly measurable mapping $\sigma : [0,T] \to G'_s$ such that

$$\sigma(s) \in F(x, X(s))$$

By definition of X and the property of ρ, we have

$$\|X(t)\| \leq \|X(0)\| + \int_0^t \|X'(s)\|ds \leq r_0 + \int_0^t \rho'(s)ds = \rho(t)$$

so that

$$\|\sigma(s)\| \leq h(s,\|X(s)\|) \leq h(s, \rho(s)) \leq \rho'(s)$$

by using (jjj) and the property of ρ. This proves that Φ is a multifunction from $\mathfrak{X}$ with non empty convex values in $\mathfrak{X}$. The remainder of the proof showing that Φ is upper semi-continuous is mutatis mutandis in the proof of theorem VI-7.

9 - For the convenience of the reader we give now a proof of Gronwall's well known inequality and its consequences which allow us to apply theorem VI-8 and to prove the global existence in cases when E is a strong dual of a separable Banach space and E is a separable Banach space.

<u>Proposition VI-9</u> - <u>Let</u> $k \in \mathcal{L}^1_+ [0,T]$ <u>and let</u> a <u>be a strictly positive number and</u> h <u>a bounded Borel mapping from</u> $[0,T]$ <u>into</u> $[0,\infty[$ <u>such that for every</u> $t \in [0,T]$,

$$h(t) \leq a + \int_0^t k(s)h(s)ds$$

<u>Then we have for every</u> $t \in [0,T]$,

$$h(t) \leq a \exp(\int_0^t k(s)ds)$$

<u>Proof</u>. Consider the function

$$\psi : t \to \operatorname{Log}(a + \int_0^t k(s)h(s)ds), \ t \in [0,T]$$

By Caratheodory's result ([3], Satz 1, p.559), ψ is absolutely continuous and we have

$$\psi'(t) = \frac{k(t)h(t)}{a + \int_0^t k(s)h(s)ds} \qquad \text{a.e.}$$

$$\leq \frac{k(t)h(t)}{h(t)} = k(t) \quad \text{for} \quad h(t) > 0$$

$$= 0 \leq k(t) \quad \text{for} \quad h(t) = 0$$

so that $\psi(t) - \psi(0) \leq \int_0^t k(s)ds$ for all $t \in [0,T]$

This implies

$$a + \int_0^t k(s)h(s)ds \leq a \exp(\int_0^t k(s)ds), \ \forall \ t \in [0,T]$$

10 - Proposition VI-10 - *Let* $h : [0,T] \times \mathbb{R} \to \mathbb{R}$ *such that for every* $t \in [0,T]$, $h(t,.)$ *is continuous on* $\mathbb{R}$ *and for every* $r \in \mathbb{R}$, $h(.,r)$ *is mesurable on* $[0,T]$. *Let* g *be a positive integrable function defined on* $[0,T]$ *such that* $|h(t,r)| \leq g(t)\ |r+1|$ *for every* $t \in [0,T]$ *and* $r \in \mathbb{R}$. *Then, for every* $r_0 \in \mathbb{R}$, *there exists* $\rho : [0,T] \to \mathbb{R}$ *such that*

$$\begin{cases} \rho(t) = r_0 + \int_0^t \rho'(s)ds \\ \rho'(t) = h(t, \rho(t)) \ \text{a.e.} \end{cases}$$

Proof. We first remark that if $\rho : [0,T] \to \mathbb{R}$ satisfies the statement of proposition VI-10, then we have

$$\rho(t) \leq |r_0| + \int_0^t g(s)|\rho(s) + 1|\ ds$$

Hence

$$\rho(t) + 1 \leq |r_0| + 1 + \int_0^t g(s)|\rho(s) + 1|\ ds$$

By Gronwall's lemma, this implies

$$\rho(t) \leq (|r_0| + 1) \exp (\int_0^t g(s)\ ds) - 1$$

Now, we prove the existence of ρ which can be deduced from theorem VI-8.

Let

$$m = (|r_0| + 1) \exp (\int_0^T g(s)ds) - 1$$

Define

$$f(t,r) = \begin{cases} h(t,r) & \text{if} \quad |r| \leq m \\ h(t,m) & \text{if} \quad r > m \\ h(t,-m) & \text{if} \quad r < -m \end{cases}$$

Consider the following equation

$$\begin{cases} \dot{z}(t) = f(t,z(t)) \\ z(0) = r_o \end{cases}$$

Then, the above equation admits at least one solution $t \to \rho(t) = r_o + \int_o^t \rho'(s)ds$ by using theorem VI-8. By virtue of Gronwall's inequality, we have $|\rho(t)| \leq m$ for all $t \in [0,T]$ so that, by definition of f,

$$\rho'(t) = h(t, \rho(t)) \quad \text{a.e.}$$

- <u>Proposition</u> VI-11 - <u>Assume that</u> h <u>satisfies the hypotheses of proposition</u> VI-10. <u>Let</u> $r \in \mathbb{R}$ <u>and define</u>

$$\mathcal{X}_r = \{z \in \mathcal{C}_{\mathbb{R}}[0,T] \,|\, z(t) = r + \int_o^t h(s,z(s))ds,\ t \in [0,T]\}$$

<u>Then there exists</u> $\tilde{z} \in \mathcal{X}_r$ <u>such that</u>

$$\tilde{z}(t) \geq z(t) \quad \text{for all } t \in [0,T] \text{ and all } z \in \mathcal{X}_r$$

<u>Proof</u> 1) Let (r_n) be a sequence in $\mathbb{R}$ such that $r = \lim_{n \to \infty} r_n$. Let (z_n) be a convergent sequence of functions belonging to $\mathcal{X}_{r_n}$. Let z be the pointwise limit of z_n. We assert that z belongs to $\mathcal{X}_r$. Observe that, for every $s \in [0,T]$,

$$\lim_{n \to \infty} h(s,z_n(s)) = h(s,z(s))$$

by continuity of $h(s,.)$. Thanks to Gronwall's inequality we can apply the Lebesgue's convergence theorem to obtain

$$z(t) = \lim_{n \to \infty} z_n(t) = r + \lim_{n \to \infty} \int_o^t h(s,z_n(s))ds$$

$$= r + \int_o^t h(s,z(s))ds$$

so that z belongs to $\mathcal{X}_r$.

2) Given two solutions z_1 and z_2 belonging to $\mathcal{X}_{r_1}$ and $\mathcal{X}_{r_2}$ respectively ; we assert now that the function

$$t \to z(t) = \sup \{z_1(t), z_2(t)\}, \quad (t \in [0,T])$$

is still a solution of $\rho'(t) = h(t, \rho(t))$. Consider the open set

$$O = \{t \in [0,T] \mid z_1(t) > z_2(t)\}$$

Then O is the countable union of its connected components ; say, $O = \bigcup_{n \in \mathbb{N}} I_n$. Define

$$y_0(t) = \begin{cases} z_1(t) & \text{if } t \in I_0 \\ z_2(t) & \text{if } t \notin I_0 \end{cases}$$

and, for $n \geq 1$,

$$y_n(t) = \begin{cases} z_1(t) & \text{if } t \in I_n \\ y_{n-1}(t) & \text{if } t \notin I_n \end{cases}$$

Then the sequence (y_n) are solutions of $\rho'(t) = h(t,\rho(t))$, hence, the pointwise limit of y_n is still a solution of $\rho'(t) = h(t,\rho(t)))$

3) Let (t_n) be a dense sequence in $[0,T]$. For every $p \geq 1$, there exists a solution $z_{n,p}$ such that

$$z_{n,p}(t_n) \geq \sup\{z(t_n) \mid z \in \mathfrak{X}_r\} - \frac{1}{p}$$

We reindex the sequence (z_n, p) by $(\eta_n)_{n \in \mathbb{N}}$. By using 2), the functions

$$y_n = \sup_{0 \leq i \leq n} \eta_i$$

belong to $\mathfrak{X}_r$ and since (y_n) is an increasing sequence, its pointwise limit $\tilde{z}$ belongs to $\mathfrak{X}_r$. Moreover we have

$\tilde{z}(t_n) = \sup\{z(t_n) \mid z \in \mathfrak{X}_r\}$ for all n so that $\tilde{z}$ satisfies the required property.

12 - <u>Proposition VI-12</u> - <u>With the hypotheses and notation of proposition</u> VI-11, <u>assume that</u> $x \to h(t,x)$ <u>is increasing for every fixed</u> $t \in [0,T]$. <u>If</u> $u \in \mathcal{C}_{\mathbb{R}}[0,T]$ <u>satisfies</u>

$$u(t) \leq r + \int_0^t h(x,u(s))ds \quad \underline{\text{for all}} \quad t \in [0,T]$$

Then we have $u(t) \leq \tilde{z}(t)$ for all $t \in [0,T]$

Proof. For every $n \geq 1$, let

$$\mathcal{X}_n = \{z \in \mathcal{C}_{\mathbb{R}}[0,T] \mid z(t) = r + \frac{1}{n} + \int_0^t h(s,z(s))ds, t \in [0,T]\}$$

By virtue of proposition VI-11, there exists $\tilde{z}_n \in \mathcal{X}_n$ such that

$$\tilde{z}_n(t) \geq z(t) \quad \text{for all} \quad t \in [0,T] \quad \text{and} \quad z \in \mathcal{X}_n$$

Moreover the function $\sup(\tilde{z}_n, \tilde{z}_{n+1})$ is a solution of $\rho'(t) = h(t,\rho(t))$ as we have already established in the proof of proposition VI-11 ; this implies that $\tilde{z}_{n+1} \leq \tilde{z}_n$ and $\tilde{z} \leq \tilde{z}_n$ for all n. Hence the pointwise limit $\tilde{z}$ of the sequence $(\tilde{z}_n)$ is still a solution of $\rho'(t) = h(t,\rho(t))$ by the first step of the proof of proposition VI-11. For every fixed n, put

$$t_0 = \sup\{t \in [0,T] \mid s \leq t \Rightarrow u(s) \leq \tilde{z}_n(s)\}$$

Since $x \to h(t,x)$ is increasing, we have

$$u(t_0) \leq r + \int_0^{t_0} h(s,u(s))ds < r + \frac{1}{n} + \int_0^{t_0} h(s, \tilde{z}_n(s))ds$$

By continuity of u and $\tilde{z}_n$, the inequality $t_0 < T$ cannot occur. So we have $u \leq \tilde{z}_n$ for all n, therefore $u \leq \tilde{z}$.

13 - We state now the global existence theorem in the case where E is a separable Banach space. We denote by E_σ the vector space E endowed with the $\sigma(E,E')$ topology.

Theorem VI-13 - Let E be a separable Banach space and let F be a multifunction from $[0,T] \times E$ with non empty convex compact values in E_σ such that

(i) for every x and every $x' \in E'$, $\delta^*(x', F(.,x))$ is measurable on $[0,T]$,

(ii) for every $t \in [0,T]$ and every $x' \in E'$, $\delta^*(x', F(t,.))$ is upper semi-continuous on E_σ

(iii) _there exists a balanced convex_ $\sigma(E,E')$ _compact set_ K _and a positive integrable function_ g _defined on_ $[0,T]$ _such that for every_ $t \in [0,T]$ _and every_ $x \in E$, $F(t,x) \subset g(t)(1 + \|x\|)K$.

Let A _be a continuous mapping from_ $[0,T]$ _into the Banach space_ $\mathcal{L}(E,E)$ _of all linear continuous mapping from_ E _into_ E. _Then, for every_ $\xi \in E$, _the set_ S_ξ _of all solutions of the multivalued differential equation_

$$(1) \quad \begin{cases} \dot{X}(t) \in A(t)\, X(t) + F(t, X(t)) \\ X(0) = \xi \end{cases}$$

is a non empty compact set in the space $\mathcal{C}_{E_\sigma}[0,T]$. _Moreover, the restriction of the multifunction_ $\xi \to S_\xi$ _to every compact set of_ E_σ _is upper-semi-continuous._

Proof. We assume without loss of generality that K is contained in the unit ball of E. Therefore, if X is a solution of the multivalued differential equation

$$(1) \quad \begin{cases} X'(t) \in A(t)\, X(t) + F(t, X(t)) \\ X(0) = \xi \end{cases}$$

We have

$$(2) \quad \|X'(t)\| \leq \|A(t)\|\, \|X(t)\| + g(t)\, (\|X(t)\| + 1)$$

Since we have

$$(3) \quad \|X(t)\| \leq \|X(0)\| + \int_0^t \|X'(s)\| ds$$

(2) and (3) imply

$$(4) \quad \|X(t)\| \leq \|X(0)\| + \int_0^t [\|A(s)\|\, \|X(s)\| + g(s)(\|X(s)\| + 1)]ds$$

Let a be a number such that $\|X(0)\| \leq a$. Then, by a simple application of Gronwall's lemma, we obtain

$$(5) \quad \|X(t)\| \leq (a+1)\, \exp\left(\int_0^t [\|A(s)\| + g(s)]ds\right)-1$$

Let us introduce now the multifunction

$$\Gamma(t) = g(t)\ (z(t) + 1)K\ , \quad t \in [0,T]$$

where $z(t) = (a+1)\ \exp(\int_0^t [\|A(s)\| + g(s)]ds) - 1$.

Since g is integrable, by virtue corollary V-4, the quotient S_Γ for the equivalence "equality a.e." of the set $\mathcal{S}_\Gamma$ of all measurable selections of Γ is non empty convex $\sigma(L^1_E\ [0,T],\ L^\infty_{E'_s}\ [0,T])$ compact.

Moreover, if X is a solution of (1), $t \to X'(t) - A(t)\ X(t)$ belongs to $\mathcal{S}_\Gamma$ since X satisfies inequality (5).

Let R be the resolvent associated with the mapping A, i.e., R is a continuous mapping from $[0,T]$ into $\mathcal{L}(E,E)$ such that

$$\begin{cases} R'(t) = A(t)\ R(t) \\ R(0) = \pi_E \end{cases}$$

where π_E is the identity mapping in E.

Let M be a compact set in E_σ. We may assume that for every $\xi \in M$, $\|\xi\| \leq a$ and let

$$\mathcal{X} = \{X : [0,T] \to E \mid X(t) = R(t)\xi + R(t) \int_0^t R^{-1}(s)f(s)ds,\ \xi \in M,\ f \in \mathcal{S}_\Gamma\}$$

Then, it is obvious that $\mathcal{X}$ is an equicontinuous subset of $\mathcal{C}_E[0,T]$, hence, a fortiori, $\mathcal{X}$ is an equicontinuous subset of $\hat{\mathcal{C}}_{E_\sigma}[0,T]$. We prove now that $\mathcal{X}$ is a relatively compact subset of $\mathcal{C}_{E_\sigma}[0,T]$. By Ascoli's theorem, it suffices to verify that, for every $t \in [0,T]$, the set

$$\mathcal{X}(t) = \{X(t) \mid X \in \mathcal{X}\}$$

is compact in E_σ. We remark first that the linear mapping $f \to \int_0^t R^{-1}(s)f(s)ds$ from $L^1_E[0,T]$ into E is continuous for every fixed $t \in [0,T]$.

Now, it is clear that $\mathcal{X}(t)$ is a compact subset of E_σ since the set

$$\mathcal{Y}(t) = \{\int_0^t R^{-1}(s)f(s)ds \mid f \in \mathcal{S}_\Gamma\}$$

is convex and $\sigma(E,E')$ compact because this set is the image of S_Γ by the weakly continuous linear mapping $f \to \int_0^t R^{-1}(s)\, f(s)\, ds$ from $L^1_E[0,T]$ into E; finally $\mathfrak{X}(t)$ is $\sigma(E,E')$ compact since we have

$$\mathfrak{X}(t) = R(t)\,(M) + R(t)\,(\mathcal{Y}(t))$$

By a standard argument already used in the proof of theorem VI-7, $\mathcal{X}$ is metrizable. Now, we can use the same arguments which are used in the proof of theorem VI-7 to finish our proof ; however the techniques of proof given here are slightly more complicated unless A is null.

First remark that the vector function

$$[0,T] \ni t \to \int_0^t R^{-1}(s)\, f(s)\, ds \quad (f \in \mathcal{S}_\Gamma)$$

is differentiable a.e. and its derivative is a.e. equal to $R^{-1}(t)\, f(t)$ by a well known result ([10], [15]) so that the vector function

$$t \to R(t)\, [\xi + \int_0^t R^{-1}(s)\, f(s)\, ds]$$

is differentiable a.e. and its derivative is a.e. equal to

$$R'(t)\, [\xi + \int_0^t R^{-1}(s)\, f(s)\, ds] + R(t)\, R^{-1}(t)\, f(t)$$

(6)

$$= A(t)\, [R(t)\xi + R(t) \int_0^t R^{-1}(s)\, f(s)\, ds] + f(t)$$

since the mapping $(U,x) \to Ux$ from $(\mathcal{L}(E,E) \times E)$ into E is bilinear and continuous.

For every $f \in \mathcal{S}_\Gamma$, define $Y_f \in \mathcal{C}_E[0,T]$ by

(7) $Y_f(t) = R(t)\xi + R(t) \int_0^t R^{-1}(s)\, f(s)\, ds$

and denote by $\Phi(f)$ the set of all measurable selections of the multi-function $t \to F(t, Y_f(t))$. Then $\Phi(f)$ is not empty by corollary VI-5 and convex.

Moreover if $f \in \Phi(f)$, Y_f is a solution of (1).

Indeed, the derivative Y'_f of Y_f defined in (7) is a.e. equal to $A\, Y_f + f$ as we have just calculated in formula (6), so

$$Y'_f(t) = A(t)\, Y_f(t) + f(t) \in A(t)\, Y_f(t) + F(t,\, Y_f(t)) \text{ a.e.}$$

Consequently we shall try to apply Kakutani-Ky Fan fixed pointxed point theorem to the multifunction Φ to obtain solutions of (1).

First, if σ is a measurable selection of $t \to F(t,\, Y_f(t))$, we have

$$\|Y'_f(t)\| \le \|A(t)\|\, \|Y_f(t)\| + \|f(t)\| \le \|A(t)\|\, \|Y_f(t)\| + g(t)\,(z(t) + 1)$$

Whence $\|Y_f(t)\| \le a + \int_0^t [\|A(s)\|\, \|Y_f(s)\| + g(s)\,(z(s) + 1)]\, ds$

so that $\|Y_f(t)\| \le z(t)$ and $\sigma(t) \in \Gamma(t)$ by proposition 12. This implies that for every $f \in S_\Gamma$, $\Phi(f) \subset S_\Gamma$. Now we prove that the graph of the multifunction Φ is compact in $(S_\Gamma)^2$. Let us observe that S_Γ is a compact metrizable space for the weak topology $\sigma(L^1_E[0,T],\, L^\infty_{E'_s}[0,T])$ since $L^1_E[0,T]$ is a separable Banach space. Let $(\sigma_n,\, f_n)$ be a sequence belonging to the graph of Φ which converges in $(S_\Gamma)^2$ to (σ, f). We have to verify that $\sigma \in \Phi(f)$.

Since the linear mapping $f \to \int_0^t R^{-1}(s) f(s)ds$ from $L^1_E[0,T]$ into E is clearly weakly continuous for every fixed $t \in [0,T]$,

$$\lim_{n \to \infty} Y_{f_n}(t) = Y_f(t)$$

for the weak topology $\sigma(E,E')$ so we can apply closure theorem VI-4. Hence $\sigma(t) \in F(t,\, Y_f(t))$ a.e. . This implies that Φ is upper-semi-continuous and admits at least a fixed point f(i.e., $f \in \Phi(f)$). To finish the proof, we have to show that the graph of the multifunction $\xi \to S_\xi$ is compact in $M \times \mathfrak{X}$. Let $(\xi_n,\, X_n)$ be a sequence belonging to the graph of S which converges to $(\xi, X) \in M \times \mathfrak{X}$. By definition of X_n, we have

$$X_n(t) = R(t)\xi + R(t) \int_0^t R^{-1}(s)\, f_n(s)ds$$

with $f_n \in S_\Gamma$ and $f_n(t) \in F(t,\, X_n(t))$ a.e. Since S_Γ is compact metrizable for the weak topology $\sigma(L^1_E,\, L^\infty_{E'_s})$ we can suppose (f_n) converges to $f \in S_\Gamma$ for this topology.

So we obtain

$$\begin{cases} X(0) = \xi \\ X(t) = R(t)\xi + \int_0^t R^{-1}(s)f(s)ds = \lim_{n\to\infty} [R(t)\xi_n + \int_0^t R^{-1}(s)f_n(s)ds] \end{cases}$$

We can again apply the closure theorem VI-4 to obtain $f(t) \in F(t, X(t))$ a.e.

Hence, X is a solution of

$$\begin{cases} X'(t) \in A(t)\, X(t) + F(t, X(t)) \\ X(0) = \xi \end{cases}$$

and clearlytely (ξ, X) belongs to the graph of S. This completes the proof.

§ 5 - SELECTION THEOREM FOR A SEPARATELY MEASURABLE AND SEPARATELY ABSOLUTELY CONTINUOUS MULTIFUNCTION.

14 - In chapter VII we shall give the global existence of solutions of a multivalued "stochastic" differential equation

$$\begin{cases} -X'(\omega,t) \in F(\omega,t, X(\omega,t)) \\ X(\omega,0) = v_o(\omega) \in \Gamma(\omega,0) \end{cases}$$

where $F(\omega,t,x)$ is the normal cone at the point x belonging to the closed convex set $\Gamma(\omega,t)$ of a separable Hilbert space H,

$$F(\omega,t,x) = \{z \in H \,|\, \forall\, u \in \Gamma(\omega,t),\ \langle z, u-x \rangle \leq 0\}$$

t underlying $[0,T]$ and ω underlying a probability space $(\Omega, \mathcal{Q}, P)$. The above problem was studied first by Moreau in several papers given in Seminar of Convex Analysis Montpellier in the deterministic case (i.e. Γ depends only on t). This kind of stochastic multivalued differential equation is entirely new since the domain of definition of the solutions depends on ω and t (in fact, under some regularity hypotheses, we shall prove the existence of a "solution whose definition will be given further). We first give the proof of a new selection theorem which will be used in the statement of the existence ofe of solutions of the multivalued stochastic differential equation

$$\begin{cases} - X'(\omega,t) \in F(\omega,t,\ X(\omega,t)) \\ X(\omega,0) = y_0(\omega) \in \Gamma(\omega,0) \end{cases}$$

Theorem VI-14 - Let $(\Omega,\mathcal{A},P)$ be a complete probability space, E a separable Banach space. Let Γ be a multifunction from $\Omega \times [0,T]$ with non empty convex $\sigma(E',E)$ closed locally compact values which contain no line in E'. Assume the following hypotheses :

(i) $L^1_E(\Omega,\mathcal{A},P)$ is separable

(ii) For every $\omega \in \Omega$ and $0 \leq \tau \leq t \leq T$,

$$h(\Gamma(\omega,t),\ \Gamma(\omega,\tau)) \leq \int_\tau^t g(s)ds$$

where g is a strictly positive ds-integrable function defined on $[0,T]$ and h is Hausdorff distance defined on the set of closed convex subsets of the strong dual E'_b.

(iii) For every $t \in [0,T]$, the multifunction $\Gamma(.,t)$ is scalarly $\mathcal{A}$-measurable, i.e., for every $x \in E$, the function $\delta^*(x\ ,\ \Gamma(.,t))$ is $\mathcal{A}$-measurable,

(iv) There exists a scalarly $\mathcal{A}$-measurable and bounded mapping y_0 from Ω into the weak dual E'_s such that $y_0(\omega) \in \Gamma(\omega,0)$ for every $\omega \in \Omega$.

Then, there exists a mapping X from $\Omega \times [0,T]$ into E' such that :

(j) for every $\omega \in \Omega$ and for every $t \in [0,T]$

$$X(\omega,t) = y_0(\omega) + \int_0^t X'(\omega,s)ds$$

where X' is a scalarly $P \otimes ds$ integrable mapping from $\Omega \times [0,T]$ into E'_s

(jj) there exists a P-negligible set N such that for every $(\omega,t) \in \Omega \backslash N \times [0,T]$, we have $X(\omega,t) \in \Gamma(\omega,t)$.

Proof. We first remark that E'_s is a Lusin space so that condition (iii) implies that for every $t \in [0,T]$, the graph of the multifunction $\Gamma(.,t)$ belongs to $\mathcal{A} \otimes \mathcal{B}(E'_s)$ where $\mathcal{B}(E'_s)$ is the Borel tribe of E'_s by using theorem III-37. Hence the multifunction $\Gamma(.,t)$ admits at least a scalarly $\mathcal{A}$-measurable selection by virtue of theorem III-22.

Let S be a subdivision of $[0,T]$

$$o = t_0 < t_1 < \ldots < t_n = T.$$

Then, by using the implicit measurable function theorem III-38 there exists a finite sequence of scalarly $\mathcal{A}$-measurable mappings $y_k (k=o,1,..n)$ associated with this subdivision such that

$$\begin{cases} \forall\, \omega \in \Omega,\ y_k(\omega) \in \Gamma(\omega,\, t_k),\ 0 \le k \le n \\ \forall\, \omega \in \Omega,\ \|y_{k-1}(\omega) - y_k(\omega)\|_{E'_b} = d(y_{k-1}(\omega),\, \Gamma(\omega,\, t_k)),\quad 1 \le k \le n \end{cases}$$

where d is the distance defined by $d(x,y) = \|x-y\|_{E'_b}$ in the strong dual E_b of E ; we note that the convex function $y \to \|x-y\|$ is inf-compact in E'_s. Let us define the mapping X_S from $\Omega \times [0,T]$ into E' by putting, for $(\omega,t) \in \Omega \times [t_{k-1},\, t_k]$,

$$X_s(\omega,t) = \frac{r(t_k) - r(t)}{r(t_k) - r(t_{k-1})}\ y_{k-1}(\omega) + \frac{r(t) - r(t_{k-1})}{r(t_k) - r(t_{k-1})}\ y_k(\omega)$$

where $r(t) = \int_0^t g(s)ds,\ t \in [0,T]$.

By definition of (y_k) and condition (ii), we have

$$\forall\, \omega \in \Omega\ ,\ \|y_{k-1}(\omega) - y_k(\omega)\|_{E'_b} \le r(t_k) - r(t_{k-1})$$

so that X_S is of the form

$$X_S(\omega,t) = y_0(\omega) + \int_0^t g(s)\ f_S(\omega,s)ds,\ t \in [0,T]$$

where f_S is a scalarly $P \otimes ds$ integrable mapping from $\Omega \times [0,T]$ with values in the unit ball B' of E'. Denote by U the unit ball of $\mathcal{L}^\infty_{E'_s}(\Omega \times [0,T],\ P \otimes ds)$. For every f in U, let

$$X_f(\omega,t) = y_o(\omega) + \int_0^t g(s)f(\omega,s)ds, \ \forall\ \omega \in \Omega,\ \forall\ t \in [0,T]$$

Let $X_f(t)$ be the partial mapping $\omega \to X_f(\omega,t)$ $(\omega \in \Omega)$ from Ω into E'_s and X_f the mapping $t \to X_f(t)$ $(t \in [0,T])$ from $[0,T]$ into $L^\infty_{E'_s}(\Omega,\mathcal{A},P)$. Now, we put on $L^\infty_{E'_s}(\Omega,\mathcal{A},P)$ the weak topology $\sigma(L^\infty_{E'_s}(\Omega,\mathcal{A},P),\ L^1_E(\Omega,\mathcal{A},P))$ and denote by $\mathcal{C}([0,T],\ L^\infty_{E'_s}(\Omega,\mathcal{A},P))$ the vector space of continuous mappings from $[0,T]$ into $L^\infty_{E'_s}(\Omega,\mathcal{A},P)$ endowed with the topology of uniform convergence. Let $\mathfrak{X} = \{X_f | f \in U\}$. We assert that $\mathfrak{X}$ is a compact metrizable subset of $\mathcal{C}([0,T],\ L^\infty_{E'_s}(\Omega,\mathcal{A},P))$. For every $h \in L^1_E(\Omega,\mathcal{A},P)$, every $f \in U$ and every $t \in [0,T]$, we have

$$(1)\quad < X_f(t),\ h > = < y_0, h > + < h \otimes \chi_{[0,t]} g, f >$$

where $\chi_{[0,t]}$ is the caracteristic function of $[0,t]$. Then by equality (1), we deduce easily that, for every fixed t in $[0,T]$, the affine mapping $f \to X_f(t)$ from U into $L^\infty_{E'_s}(\Omega,\mathcal{A},P)$ is weakly continuous ; therefore, the set $\mathfrak{X}(t) = \{X_f(t) | f \in U\}$ is a convex compact metrizable set of $L^\infty_{E'_s}(\Omega,\mathcal{A},P)$. Moreover, it follows from equality (1), that

$$(2)\quad < X_f(t) - X_f(\tau),\ h > \leq \left(\int_\tau^t g(s)ds\right) N_1(h)$$

for $0 \leq \tau \leq t \leq T$. By virtue of Ascoli's theorem, we conclude that the set

$$\mathfrak{X} = \{Y_f | f \in U\}$$

is a relatively compact metrizable set of $\mathcal{C}([0,T],\ L^\infty_{E'_s}(\Omega,\mathcal{A},P))$. We consider now a sequence (S_n) of subdivisions of $[0,T]$

$$0 = t^n_o < t^n_1 < \ldots < t^n_i < t^n_{k_n} = T$$

such that

$$\lim_{n \to \infty} \max\{t^n_{i+1} - t^n_i | 0 \leq i \leq k_n - 1\} = 0$$

and denote by X_n the mapping associated with S_n by the above construction. Then the sequence (X_n) belongs to the convex compact metrizable set $\mathfrak{X}$, namely

$\mathcal{X} = \{Y_f \in \mathcal{C}[0,T],\ L^{\infty}_{E'_s}(\Omega,\mathcal{A},P) \mid Y_f(\omega,t) = y_0(\omega) + \int_0^t g(s)f(\omega,s)ds,\ f \in U\}$

So, there exists a subsequence $(X_{\varphi(n)})$ which converges to an element belonging to $\mathcal{X}$; namely there exists $f \in U$ such that

$$\lim_{n \to \infty} X_{\varphi(n)} = X$$

with

$$X(\omega,t) = y_0(\omega) + \int_0^t g(s)\ f(\omega,s)ds,\ \forall\ \omega \in \Omega\ ,\ \forall\ t \in [0,T]$$

To simplify our notation we can suppose that the initial sequence (X_n) converges to X. We shall prove that X_f satisfies the assertion (jj). By the construction of (X_n), there exists a sequence (θ_n) of step functions from $[0,T]$ into $[0,T]$ such that

$$\forall\ t \in [0,T],\ \lim_{n \to \infty} \theta_n(t) = t$$

$\forall\ \omega \in \Omega$, $\forall t \in [0,T]$, $X_n(\omega,\ \theta_n(t)) \in \Gamma(\omega,\ \theta_n(t))$

Let t be a fixed number in $[0,T]$, we first prove that there exists a P-negligible set $N(t)$ such that

$$X(\omega,t) \in \Gamma(\omega,t)\quad ,\ \forall\ \omega \notin N(t)$$

For every fixed ω in Ω and every fixed $t \in [0,T]$, $\Gamma(\omega,t)$ is convex $\sigma(E',E)$ closed locally compact and contains no **lines**; there exists by lemma III-34, a sequence (e_m) in E <u>which does not depend</u> on (ω,t) such that

$$\Gamma(\omega,t) = \bigcap_{m=1}^{\infty} H_m(\omega,t) .$$

with $H_m(\omega,t) = \{x' \in E' \mid < x',e_m > \leq \delta^*(\omega,t,e_m)\}$

$$\delta^*(\omega,t,e_m) = \sup_{y \in \Gamma(\omega,t)} < e_m,y > < +\infty$$

Then $X(\omega,t)$ belongs to $\Gamma(\omega,t)$ if and only if

$$< e_m,\ X(\omega,t) > \leq \delta^*(\omega,t,e_m),\ \forall\ m$$

Consider the $\mathcal{A}$-measurable set,

$$A = \{\omega \in \Omega \mid X(\omega,t) \notin \Gamma(\omega,t)\}$$

and suppose $P(A) > 0$. Then, there exists at least an element e_m and $\mathcal{Q}$-measurable set A_m,

$$A_m = \{\omega \in \Omega | < e_m, X(\omega,t) >> \delta^*(\omega,t,e_m)\}$$

such that $P(A_m) > 0$. Hence, by integrating on A_m, we obtain

$$(1) \quad \int_{A_m} < e_m, X(\omega,t) > P(d\omega) > \int_{A_m} \delta^*(\omega,t,e_m) P(d\omega)$$

But we have, for every n, every ω and every t,

$$< e_m, X_n(\omega,\theta_n(t)) > \leq \delta^*(\omega,\theta_n(t), e_m)$$

It follows by integrating on A_m, that

$$(2) \quad \int_{A_m} < e_m, X_n(\omega,\theta_n(t)) > P(d\omega) \leq \int_{A_m} \delta^*(\omega,\theta_n(t),e_m) P(d\omega)$$

Since (X_n) converges to X in the compact metrizable set $\mathcal{X}$ as $n \to +\infty$, we have

$$\lim_{n \to \infty} \int_{A_m} < e_m, X_n(\omega,\theta_n(t)) > P(d\omega) = \int_{A_m} < e_m, X(\omega,t) > P(d\omega)$$

By condition (ii) we have

$$|\delta^*(\omega,\theta_n(t),e_m) - \delta^*(\omega,t,e_m)| \leq \|e_m\| \, |r(\theta_n(t)) - r(t)|$$

This implies

$$\lim_{n \to \infty} \int_{A_m} \delta^*(\omega,\theta_n(t),e_m) P(d\omega) = \int_{A_m} \delta^*(\omega,t,e_m) P(d\omega)$$

Hence, we deduce from inequality (2)

$$\int_{A_m} < e_m, X(\omega,t) > P(d\omega) \leq \int_{A_m} \delta^*(\omega,t,e_m) P(d\omega)$$

This contradicts inequality (1). Now, by using the continuity of functions $\delta^*(\omega,.,e_m)$ on $[0,T]$ we conclude easily that there exists a P-negligible N which <u>does not depend</u> on $t \in [0,T]$ such that

$$X(\omega,t) \in \Gamma(\omega,t) \quad , \quad \forall \omega \notin N, \ \forall t \in [0,T]$$

That completes the proof.

BIBLIOGRAPHY OF CHAPTER VI

1 - ARTSTEIN, Z. : Calculus of compact set-valued functions. Research program in game theory and mathematical economics n°57. The Hebrew university of Jerusalem 1970.

2 - BERGE, C. : Espaces topologiques. Fonctions multivoques. Dunod 1959.

3 - CARATHEODORY, C. : Vorlesnungen über reelle funktionen Chelsea 1968.

4 - CASTAING, Ch. : Sur les équations différentielles multivoques. C.R. Acad. Sc. Paris 264, 63-66 (1966).

5 - CASTAING, Ch. : Un résultat de dérivation des multi-applications. Exposé n°2. Séminaire d'Analyse convexe Montpellier 1974.

6 - CASTAING, Ch. : Un théorème d'existence de sections séparément mesurables et séparément absolument continues. C.R. Acad. Sc. Paris 276, 367-370 (1973) et Exposé n°3 Séminaire d'Analyse convexe Montpellier 1973.

7 - CASTAING, Ch., VALADIER, M. : Equations différentielles multivoques dans les espaces vectoriels localement convexes. Revue Informatique et de Recherche Operationelles 16, 3-16 (1969).

8 - DAURES, J.P. : Contribution à l'étude des équations différentielles multivoques dans les espaces de Banach C.R. Acad. Sc. Paris 270, 769 769-772 (1970).

9 - DINCULEANU, N. : Vector measures Pergamon Press 1967.

10 - DUNFORD, N., SCHWARTZ, J.T : Linear operators, Interscience, New York 1958.

11 - GHOUILA HOURI, A. : Equations différentielles multivoques C.R. Acad. Sc. Paris 261, 2568-2571 (1965).

12 - KOMURA, Y. : Non linear semi-groups in Hilbert spaces. Journal Math. Soc. Japan 19, 493-507 (1967).

13 - LASOTA, A., OPIAL, Z. : An application of the Kakutani-Ky Fan theorem in the theory of ordinary differential equation. Bull. Acad. Pol. Sc. Math 13,781-786 (1965)

14 - VALADIER, M. : Existence globale pour les équations différentielles multivoques. C.R. Acad. Sc. Paris 272, 274-477 (1971) et Exposé n°11 Séminaire d'Analyse Convexe Montpellier 1971.

15 - YOSIDA, K. : Functional analysis, 2nd ed. Berlin Springer Verlag 1968.

CHAPTER VII

CONVEX INTEGRAND ON LOCALLY CONVEX SPACES. AND ITS APPLICATIONS.

In this section we describe the theory of convex functionals of the form

$$I_f(u) = \int_T f(t, u(t))\ \mu(dt)$$

which was first studied by Rockafellar in many papers ([20], [21], [22]). The main theorems extend a theorem of Rockafellar which was stated for separable reflexive Banach spaces. We first give below some technical results of measurability in order to characterize the normal convex integrands. Second, Rockafellar's definition of decomposable spaces of vector valued functions is extended to locally convex spaces allowing us to generalize Rockefellar's duality theorem between a pair od decomposable spaces of vector valued functions. We deduce from the above results Strassen's celebrated theorem ([23]), a disintegration theorem ; finally we shall use the duality theorem to prove the existence of a solution of a stochastic evolution equation introduced in paragraph 14 of chapter VI.

In everything that follows, let $(T, \mathcal{T}, \mu)$ be a σ-finite positive measure space with $\mathcal{T}$ μ-complete. Let E be a locally convex space. We assume that E and its dual E' are Suslin locally convex spaces for topologies which are compatible with duality. For a topological space F, we denote by $\mathcal{B}(F)$ the Borel tribe of F.

§ 1 - PRELIMINARY RESULTS OF MEASURABILITY.

Definition. A mapping $f : T \times E \to]-\infty, +\infty]$ is called a normal integrand if for every fixed t in T, $f(t,.)$ is lower-semi-continuous and f is

$\mathcal{C}\otimes\mathcal{B}(E)$-measurable ; f is called a convex normal integrand if it is a normal integrand and for every fixed t in T, f(t,.) is convex.

1 - Lemma VII-1 - Let f be a mapping from $T \times E$ to $]-\infty, +\infty]$. Then f is a normal integrand if and only if for every fixed t in T

$$\text{epi } f_t = \{(x,r)\in E\times\mathbb{R} \mid f(t,x)\leq r\}$$

is closed and the graph of the multifunction $t \to \text{epi } f_t$ belongs to $\mathcal{C}\otimes\mathcal{B}(E\times\mathbb{R})$.

Proof. First, f(t,.) is lower-semi-continuous if and only if epi f_t is closed in the product space $E\times\mathbb{R}$.

1) If f is a normal integrand. Then the graph G of $t \to \text{epi } f_t$ given by the formula

$$G=\{(t,x,r)\in T\times E\times\mathbb{R} \mid f(t,x)\leq r\}$$

belongs to $\mathcal{C}\otimes\mathcal{B}(E)\otimes\mathcal{B}(\mathbb{R})$ which is equal to $\mathcal{C}\otimes\mathcal{B}(E\times\mathbb{R})$ (since it is easy to prove the equality $\mathcal{B}(E)\otimes\mathcal{B}(\mathbb{R}) = \mathcal{B}(E\times\mathbb{R})$ using the fact that $\mathbb{R}$ has a countable basis of open sets).

2) Conversely, assume that the graph G of $t \to \text{epi } f_t$ belongs to $(\mathcal{C}\otimes\mathcal{B}(E))\otimes\mathcal{B}(\mathbb{R})$. Then, for every fixed r in $\mathbb{R}$, the set

$$\{(t,x)\in T\times E \mid (t,x,r)\in G\}$$

belongs to $\mathcal{C}\otimes\mathcal{B}(E)$ (see Neveu, prop. III-1-2)

But

$$\{(t,x)\in T\times E \mid (t,x,r)\in G\} = \{(t,x)\in T\times E \mid f(t,x)\leq r\}$$

Thus f is $\mathcal{C}\otimes\mathcal{B}(E)$-measurable.

2 - Corollary VII-2 - If f is a normal integrand on $T\times E$, then the function

$f^* : T\times E' \to]-\infty, +\infty]$

$$f^*(t,x') = \sup\{\langle x',x\rangle - f(t,x) \mid x\in E\}$$

is a convex normal integrand on $T\times E'$.

Proof. It is obvious from the definition of f^* that for every $t \in T$ and x' in E',

$$f^*(t,x') = \sup\{\langle x',x\rangle - r \mid (x,r) \in \text{epi } f_t\}$$

Since f is a normal integrand, by lemma VII-1, the graph of the multi-function, $t \to \text{epi } f_t$ belongs to $\mathcal{T} \otimes \mathcal{B}(E \times \mathbb{R})$. So, by virtue of theorem III-22 there exists a sequence of measurable selections (u_n, r_n) of $t \to \text{epi } f_t$ such that for every fixed t in T, the sequence $(u_n(t), r_n(t))$ is dense in epi f_t. Thus, we have

$$f^*(t,x') = \sup_n [\langle x', u_n(t)\rangle - r_n(t)]$$

Hence f^* is $\mathcal{T} \otimes \mathcal{B}(E')$-measurable since $(t,x') \to \langle x', u_n(t)\rangle$ is $\mathcal{T} \otimes \mathcal{B}(E')$ measurable for every n by applying again theorem III-22.

3 - Definition. Let $\mathcal{L}_E$ (resp. $\mathcal{L}_{E'}$) be a vector space of scalarly integrable functions from T to E (resp. E') such that for every u (resp. v) belonging to $\mathcal{L}_E$ (resp. $\mathcal{L}_{E'}$) the scalar function $t \to \langle u(t), v(t)\rangle$ is integrable on T (remark that this function is automatically measurable by virtue of theorem III-22. Let $\mathcal{M}_E$ (resp. $\mathcal{M}_{E'}$) be the vector space of scalarly measurable functions f from T to E(resp. E') such that $f(T)$ is relatively compact in E (resp. E'). The space $\mathcal{L}_E$ (or similarly $\mathcal{L}_{E'}$) is said to be decomposable if, whenever u belongs to $\mathcal{L}_E$ and f belongs to $\mathcal{M}_E$, A belongs to $\mathcal{T}$ with finite measure, the function $\chi_A f + \chi_{T\setminus A} u$ also belongs to $\mathcal{L}_E$.

Remark. Let E be a separable reflexive Banach space and let E_σ be the vector space E endowed with the weak topology $\sigma(E,E')$, then E_σ is a Lusin locally convex space because E is a Polish space and the topology of E is finer than the weak topology $\sigma(E,E')$. So, definition 3 is equivalent to Rockafellar's since $\mathcal{L}^\infty_{E_\sigma}$ is $\mathcal{M}_{E_\sigma}$.

Example. As an obvious example, one could take $F = E'_s$ where E'_s is the dual of the separable Frechet space E and E'_s is equipped with the weak topology $\sigma(E',E)$. Indeed, E'_s is a Lusin locally convex space (because if (V_n) is a countable basis of neighbourhoods of the origin in E, the polar V°_n of V_n is a compact metrizable set in E'_s and E'_s is equal to the countable union of V^n_o, that is $E'_s = \bigcup_n V^o_n$). So $\mathcal{L}^\infty_F = \mathcal{M}_F$ is decomposable and it is obvious that the vector space $\mathcal{L}^1_E$ of all absolutely summable functions from T to E is also decomposable.

4 - The following propositions will be used to state our main theorem.

Proposition VII-4 - Let E be a locally convex Suslin space and let u be a scalarly measurable mapping from T to E. Then there exists a sequence (T_n) of $\mathcal{C}$-measurable sets such that

$$\begin{cases} T_n \cap T_m = \emptyset \text{ if } n \neq m \\ \mu(T - \bigcup_n T_n) = 0 \\ \forall n, \overline{u(T_n)} \text{ is compact.} \end{cases}$$

Proof. Since μ is σ finite, it is sufficient to prove the proposition when μ is bounded. By theorem III-36, u is a Borel mapping, that is, for every Borel set B in E, $u^{-1}(B)$ belongs to $\mathcal{C}$. So, we can consider the Borel measure ν defined on E by

$$\nu(B) = \mu[u^{-1}(B)], \ \forall B \in \mathcal{B}(E)$$

But E is Suslin, ν is a Radon measure (see Bourbaki ([3]), prop 3, p.49). Hence, there exists a sequence (K_n) of compact sets in E such that

$$\begin{cases} K_n \cap K_m = \emptyset, \ n \neq m \\ \nu(\bigcup_n K_n) = \nu(E) \end{cases}$$

Now, it is easy to check that the sequence (T_n) defined by $T_n = u^{-1}(K_n)$ satisfies the expected properties.

5 - <u>Proposition VII-5</u> - <u>Assume that</u> $\mathcal{L}_E$ <u>is decomposable</u>. <u>Let</u> v <u>be an element of</u> $\mathcal{L}_{E'}$ <u>such that</u>

$$< v,u > = \int_T < v(t),\ u(t) > \mu(dt) = 0$$

<u>for all</u> u <u>belonging to</u> $\mathcal{L}_E$. <u>Then</u> v <u>is equal to zero almost everywhere</u>.

<u>Proof</u>. Let A be an arbitrary $\mathcal{C}$ measurable set in T with $\mu(A) < +\infty$. Note that if x is an element in E and φ is a real valued bounded measurable function, then φx belongs to $\mathcal{U}_E$. But $\mathcal{L}_E$ is decomposable, $\chi_A\ \varphi x$ belongs to $\mathcal{L}_E$. This implies

$$\int_A < v(t),\ x > \varphi(t)\ \mu\ (dt) = 0$$

Hence $< v(t),x > = 0$ almost everywhere. As E is the dual of E', there exists in E a sequence (e_n) which separates points of E by virtue of lemma III-31. Hence $v(t) = 0$ almost everywhere.

6 - <u>Corollary VII-6</u> - <u>Let</u> L_E (resp. $L_{E'}$) <u>be the vector space of equivalence classes under equality almost everywhere</u>. <u>If</u> $\mathcal{L}_E$ <u>and</u> $\mathcal{L}_{E'}$ <u>are decomposable, then the bilinear mapping</u> $(u,v) \to < u,v >$ <u>defines a separated duality between</u> L_E <u>and</u> $L_{E'}$.

<u>Proof</u>. This follows directly from proposition VII-5.

2 - <u>DUALITY THEOREM OF INTEGRAL CONVEX FUNCTIONALS FOR LOCALLY CONVEX SUSLIN SPACES.</u>

7 - <u>Convention</u>. If f is a normal integrand and if u belongs to $\mathcal{L}_E$, we adopt the following convention

$$I_f(u) = \begin{cases} \int_T f(t,u(t)\mu(dt) & \text{if } \int_T f(t,u(t))^+\mu(dt) < +\infty \\ +\infty & \text{if } \int_T f(t,u(t))^+\mu(dt) = +\infty \end{cases}$$

<u>and similarly for the conjugate</u> f^* of f (See corollary VII-2).

Theorem VII-7 - If $\mathcal{L}_E$ is decomposable and if the integral functional I_f defined above on $\mathcal{L}_E$ is finite for at least one u_o in $\mathcal{L}_E$, then the convex integral functional I_{f*} on $\mathcal{L}_{E'}$

$$I_{f*}(v) = \begin{cases} \int_T f^*(t,v(t))\mu(dt) & \text{if } \int_T f^*(t,v(t))^+\mu(dt) < +\infty \\ +\infty & \text{if } \int_T f^*(t,v(t))\,\mu(dt) = +\infty \end{cases}$$

is conjugate to the integral functional I_f on $\mathcal{L}_E$, that is,

$$I_{f*}(v) = \sup\{< v,u > - \ I_f(u) \,|\, u \in \mathcal{L}_E\}$$

for every v in $\mathcal{L}_{E'}$.

If, in addition f_t is convex, $\mathcal{L}_E$ is decomposable and $I_{f*}(v) < +\infty$ for at least one $v_o \in \mathcal{L}_{E'}$, then I_f and I_{f*} are proper convex lower-semi-continuous conjugate to each other.

Proof. We need only prove that for every v in $\mathcal{L}_{E'}$,

$$\int_T f^*(t,v(t))\mu(dt) \leq \sup\{< v,u > - I_f(u) \,|\, u \in \mathcal{L}_E\}$$

since the inequality $\geq$ is obvious. Let β be a real number such that $\beta < I_{f*}(v)$. We have to prove the existence of a function $u \in L_E$ such that

$$< u,v > - I_f(u) \geq \beta$$

thereby establishing the theorem. Since $I_f(u_o)$ is finite, there exists a real integrable function a_o such that

$$< v(t),\ u_o(t) > - f(t,\ u_o(t)) \geq a_o(t)$$

for all $t \in T$. This implies that

$$f^*(t,\ v(t)) \geq a_o(t)$$

for all $t \in T$. We assert now the existence of a real integrable function γ such that

$$\begin{cases} \int_T \gamma(t)\mu(dt) > \beta \\ \gamma(t) < f^*(t,v(t)) \quad \text{for all } t \in T. \end{cases}$$

As μ is σ-finite, there exists a strictly positive integrable function h defined on T. If $I_{f^*}(v)$ is finite, we take

$$\gamma(t) = f^*(t,v(t)) - \varepsilon\, h(t)$$

where ε is a sufficiently small strictly positive number. If $I_{f^*}(v)$ is $+\infty$, consider the sequence of integrable functions (ξ_n) on T

$$\xi_n(t) = \begin{cases} \inf\,[n\, h(t), \frac{1}{2} f^*(t,v(t))] \text{ if } f^*(t,v(t)) > 0 \\ f^*(t,v(t)) - h(t) \text{ if } f^*(t,v(t)) \le 0 \end{cases}$$

By the definition of (ξ_n), we have

$$\lim_{n \to \infty} \xi_n(t) = \frac{1}{2} f^*(t,v(t)) \text{ if } f^*(t,v(t)) > 0$$

so that $\lim_{n \to \infty} \int \xi_n(t)\mu(dt) = +\infty$ by monotone convergence theorem.

So there exists an integer N_0 such that

$$\int \xi_{N_0}(t)\mu(dt) > \beta$$

Thus it suffices to take $\gamma(t) = \xi_{N_0}(t)$ $(t \in T)$ to obtain the inequality $\gamma(t) < f^*(t,v(t))$ for all $t \in T$. Let us consider now the multifunction Γ

$$\Gamma(t) = \{x \in E \,|< v(t),x > - f(t,x) \ge \gamma(t)\}, \quad \forall\, t \in T$$

Then Γ is a multifunction from T with non empty closed values in E such that the graph $G(\Gamma)$ of Γ,

$$G(\Gamma) = \{(t,x) \in T \times X \mid < v(t),x > - f(t,x) \ge \gamma(t)\}$$

belongs to $\mathcal{T} \otimes \mathcal{B}(E)$ so that Γ admits a measurable selection s by virtue of theorem III-22.

By proposition VII-4, we can find an increasing sequence (T_n) in $\mathcal{T}$ such that

$$\begin{cases} \mu(T_n) < +\infty \\ \mu(T \setminus \bigcup_n T_n) = 0 \\ \overline{s(T_n)} \text{ is compact.} \end{cases}$$

For each n let

$$u_n(t) = \begin{cases} s(t) & \text{if } t \in T_n \\ u_0(t) & \text{if } t \in T\setminus T_n \end{cases}$$

Then u_n belongs to $\mathcal{L}_E$ by the decomposability hypothesis . For each n we have

$$< v(t), u_n(t) > - f(t, u_n(t)) \geq \gamma(t) \quad \text{if} \quad t \in T_n$$

$$< v(t), u_n(t) > - f(t, u_n(t)) \geq \alpha(t) \quad \text{if} \quad t \in T\setminus T_n$$

Thus

$$\int_T < v(t), u_n(t) > \mu(dt) - \int_T f(t,u_n(t))\mu(dt) \geq \int_{T_n} \gamma(t)\mu(dt)+\int_{T\setminus T_n} \alpha(t)\mu(dt)\geq\beta$$

when n is large ; so we can conclude that

$$I_{f^*}(v) = \sup\{< v,u > - I_f(u) \mid u \in \mathcal{L}_E\}$$

<u>Remark</u>. When E is a separable Banach space the notion of decomposable vector space given in definition 3 can be developed by introducing the vector space $\mathcal{M}_E$ (resp. $\mathcal{M}_{E'_s}$) of scalarly measurable functions f from T to E (resp. E'_s) such that $f(T)$ is <u>bounded</u> in E (resp. E'_s) and theorem VII-7 still holds in this case. Indeed, we have already remarked that E and E'_s are Lusin spaces. Moreover if v is a scalarly measurable mapping from T to E (resp. E'_s), then the scalar function $t \to \|v(t)\|_E$ (resp. $t \to \|v(t)\|_{E'_s}$) where $\|.\|_E$ (resp. $\|.\|_{E'_b}$) is the norm in the Banach space E (resp. strong dual E'_b of E) is measurable by noting that

$$\begin{cases} \|v(t)\|_E = \sup\limits_{\|x'\|\leq 1} < v(t), x' > \\ \text{resp. } \|v(t)\|_{E'_b} = \sup\limits_{\|x\|\leq 1} < v(t),x > \end{cases}$$

so, it is very easy to choose an increasing sequence of $\mathcal{T}$-measurable sets T_n of finite measure with union T such that $v(T_n)$ is bounded. Thus the arguments given in the proof of theorem VII-7 are applicable without using the proposition VII-4 in which the basic tool is Bourbaki's result on Suslin spaces ([3]).

§ 3 - DUALITY THEOREM OF CONVEX INTEGRAL FUNCTIONALS IN NOW SEPARABLE REFLEXIVE BANACH SPACE.

8 - In the following we give an analogous version of theorem VII-7. Namely we state a duality theorem in the special case where E is a reflexive Banach space (without assuming the separability hypothesis on E) which will be used to state the existence of solutions of certain classes of evolution equations. However we need here some topological hypotheses on T. We suppose below that T is a compact space equipped with a positive Radon measure μ. This enables us to use the theory of measurable functions given in Bourbaki's books that allows us to give the definition of a decomposable vector space of μ-measurable functions from T with values in a reflexive Banach space without assuming the separability hypothesis. We first give below two elementary measurability results which are very useful for our purpose. We denote by E a reflexive Banach space and E'_b the strong dual of E.

Lemma VII-8 - Let E_σ be the vector space E equipped with the weak topology $\sigma(E, E')$ and Z a Hausdorff topological space. Let h be a real function defined on $Z \times E_\sigma$ taking its values in $]-\infty, +\infty]$ with the following property :

For each real number λ, the set

$$A_\lambda = \{(t,\alpha) \in Z \times E_\sigma \mid h(t,x) > \lambda\}$$

is a countable union of compact sets in $Z \times E_\sigma$. Let Δ be a multifunction from Z with non empty closed values in E_σ such that the graph $G(\Delta)$ of Δ,

$$G(\Delta) = \{(t,x) \in Z \times E_\sigma \mid x \in \Delta(t)\}$$

is closed in $Z \times E_\sigma$. Then the function

$$\rho : t \mapsto \sup\{h(t,x) \mid x \in \Delta(t)\} \quad (t \in Z)$$

is Borel measurable on Z.

Proof. For each real number λ, we have

$$\{t \in Z \mid \rho(t) > \lambda\} = \mathrm{proj}_Z\, [G(\Delta) \cap A_\lambda]$$

with

$$A_\lambda = \{(t,x) \in Z \times E_\sigma \mid h(t,x) > \lambda\}$$

Since A_λ is a countable union of compact sets in $Z \times E_\sigma$, $G(\Delta) \cap A_\lambda$ has the same property. Thus, $\mathrm{proj}_Z[G(\Delta) \cap A_\lambda]$ is a countable union of compact sets, so we conclude that ρ is Borel measurable on Z.

Remarks 1) If Z is $\mathbb{R}$, we can suppose that the graph $G(\Delta)$ of Δ is sequentially closed since $\mathbb{R} \times E$ is a reflexive Banach space.

2) This lemma still holds if we replace E_σ by a Hausdorff topological space which is a countable union of compact spaces.

9 - Lemma VII-9 - Let f be a lower-semi-continuous function from $T \times E_\sigma$ to $]-\infty, +\infty]$. Let v be a μ-measurable mapping from T to E'_s. Then, the function

$$t \to f^*(t, v(t)) = \sup\{\langle x, v(t)\rangle - f(t,x) \mid x \in E\}$$

is μ-measurable on T.

Proof. It follows from the definition of $f^*(t,.)$ that

$$f^*(t,v(t)) = \sup\{\langle x, v(t)\rangle - r \mid (x,r) \in \mathrm{epi}\ f_t\}$$

But f is lower semi-continuous, so the graph G of the multifunction $t \to \mathrm{epi}\ f_t$

$$G = \{(t,x,r) \in T \times E_\sigma \times \mathbb{R} \mid f(t,x) \leq r\}$$

is closed. Since v is a weakly μ-measurable mapping, v is a strongly μ-measurable mapping by the Dunford-Pettis-Phillips theorem ([7], p. 314).

Hence, for every $\varepsilon > 0$, there exists a compact set $K_\varepsilon \subset T$ with $\mu(T \setminus K_\varepsilon) \leq \varepsilon$ such that the function v is continuous from K_ε to the strong dual E'_s of E. Let B be the unit ball of E_σ. Then the mapping $h : (t,x,r) \to \langle x, v(t) \rangle - r$ from $K_\varepsilon \times \rho B \times \mathbb{R}$ to $\mathbb{R}$ is continuous for every positive number ρ when ρB is equipped with the $\sigma(E,E')$ topology. So for every λ in $\mathbb{R}$, the set

$$A_\lambda = \{(t,x,r) \in K_\varepsilon \times E_\sigma \times \mathbb{R} \mid h(t,x,r) > \lambda\}$$

$$= \bigcup_n \{(t,x,r) \in K_\varepsilon \times nB \times R \mid h(t,x,r) > \lambda\}$$

is a countable union of compact sets in $K_\varepsilon \times E_\sigma \times \mathbb{R}$. So we deduce from lemma VII-8 and remark 2) of this lemma by taking $Z = K_\varepsilon$, $\Delta(t) = \text{epi } f_t$ that the function $t \mapsto f^*(t,v(t))$ is Borel measurable on K_ε. Since ε is arbitrary, this proves that $t \mapsto f^*(t,v(t))$ is μ-measurable on T.

10 - In order to state a duality theorem for reflexive Banach spaces, we need a theorem of measurable selection of a multifunction from a compact space Z with non empty convex closed values in the weak space E_σ of a reflexive Banach space E such that its graph is closed in $Z \times E_\sigma$. However, we will prove below this result in a more general setting. Namely, we introduce the property of <u>pseudo-upper semi-continuity</u> of a multifunction that follows.

<u>Definition</u>. Let X be a topological space, G be a Hausdorff topological space. A multifunction Γ from W with values in G is said to be <u>pseudo-upper-semi-continuous</u> on X if for every compact set K in G, the multifunction

$$\Sigma_K : t \mapsto \Gamma(t) \cap K \quad (t \in X)$$

is upper-semi-continuous on X, that is, for every closed set F in G,

$$\Sigma_K^{-}F = \{t \in X \mid \Sigma_K(t) \cap F \neq \emptyset\}$$

is closed.

Examples

1 - *If* Γ *is a multifunction from* X *with values in* G *such that the graph of* Γ *is closed, then* Γ *is pseudo-upper-semi-continuous on* X *(see for instance Berge* [1], p.119).

2 - *If* X *is a metrizable space, if, for every relatively compact sequence in* G *one can extract a converging subsequence and if* Γ *has the following property : For every* t_0 *in* X *and for every sequence* (t_n) *which converges to* t_0, *one has*

$$\bigcap_k \overline{\bigcup_{n\geq k} \Gamma(t_n)} \subset \Gamma(t_0)$$

then, Γ *is pseudo-upper-semi-continuous on* X.

Let K be a compact set in G. We have to prove that, for every closed set F in G,

$$\Sigma_K^- F = \{t \in X \mid \Gamma(t) \cap K \cap F \neq \emptyset\}$$

is closed. Let (t_n) be a sequence of points belonging to $\Sigma_K^- F$ which converges to t_0. Let u_n be an element belonging to $\Gamma(t_n) \cap K \cap F$. Since $K \cap F$ is compact, the sequence (u_n) is relatively compact. So we can extract from (u_n) a subsequence which converges to $u_0 \in K \cap F$. We still denote this subsequence by (u_n).
Thus, we have

$$u_0 \in \bigcap_k \overline{\bigcup_{n\geq k} \{u_n\}} \subset \bigcap_k \overline{\bigcup_{n\geq k} \Gamma(t_n)} \subset \Gamma(t_0)$$

This proves that u_0 belongs to $\Gamma(t_0) \cap K \cap F$, that is, $\Sigma_K^- F$ is closed.

3 - *If* X *is metrizable, if* G *is a Hausdorff locally convex space such that from every relatively compact sequence in* G *one can extract a converging subsequence, and if* Γ *has the following property : For every* t_0 *in* X *and for every sequence* (t_n) *which converges to* t_0, *one has*

$$\bigcap_k \overline{co\left(\bigcup_{n\geq k} \Gamma(t_n)\right)} \subset \Gamma(t_0)$$

where $\overline{co}$ *is the closed convex hull, then* Γ *is pseudo-upper-semi-continuous on* X.

The proof of example 3 is left to the reader since one can use the arguments given in example 2 because this example is a particular case of example 2.

- *Lemma VII-11* - *Let* X *be a Hausdorff topological space, and* G *be a Hausdorff topological space. Let* f *be a mapping from* $X \times G$ *into* $[-\infty,+\infty]$ *with the following properties* :

a) $$\inf\{f(t,x) \mid (t,x) \in X \times G\} > -\infty$$

b) *For every compact set* K *in* G, f *is lower-semi-continuous on* $X \times K$,

c) *The set* $\mathrm{dom}\ f_t = \{x \in G \mid f(t,x) < +\infty\}$ *is closed and not empty for every* t *in* X

d) *The multifunction* $t \mapsto \mathrm{dom}\ f_t$ *is pseudo-upper-semi-continuous on* X.

Let g *be an inf-compact mapping from* G *into* $\mathbb{R}$. *Then the function*

$$t \to \varphi(t) = \inf_{u \in G} \{f(t,u) + g(u)\} \qquad (t \in X)$$

is lower-semi-continuous on X. *Moreover, if* μ *is a bounded positive Radon measure on* X, *then for every* $\epsilon > 0$, *there exists a compact set* $X_0 \subset X$ *with* $\mu(X \setminus X_0) \leq \epsilon$ *such that the restriction to* X_0 *of the multifunction*

$$t \to \Phi(t) = \{x \in G \mid f(t,x) + g(x) \leq \varphi(t)\}$$

is upper-semi-continuous.

Proof. We shall use here some arguments given by Moreau ([14], p.23). It is clear that, for every t in X, we have

$$\varphi(t) = \inf\{f(t,u) + g(u) \mid u \in \mathrm{dom}\ f_t\}$$

For every fixed t_0 in X and every fixed a in $\mathbb{R}$ such that $\varphi(t_0) > a$, we have to prove the existence of a neighbourhood V of t_0 in X such that $\varphi(t) \geq a$ for all t in V. Let β be a minorant of f. Then, the set

$$K = \{u \in G \mid g(u) \leq a - \beta\}$$

is compact because g is inf-compact. But we have, for every t in X

$$\varphi(t) = \inf\{\varphi_1(t), \varphi_2(t)\}$$

with

$$\varphi_1(t) = \inf\{f(t,u) + g(u) \mid u \in \text{dom } f_t \setminus K\} \geq a$$

$$\varphi_2(t) = \inf\{f(t,u) + g(u) \mid u \in \text{dom } f_t \cap K\}$$

Since the function $(t,u) \mapsto f(t,u) + g(u)$ is lower semi-continuous on $X \times K$ and the multifunction $t \mapsto \text{dom } f_t \cap K$ is upper semi-continuous on X with compact values in G, φ_2 is lower-semi-continuous on X by virtue of a result due to Berge ([1], p. 115). But we have $\varphi_2(t_0) > a$, hence there exists a neighbourhood V of t_0 in X such that $\varphi_2(t) \geq a$ for all t in V ; this proves that φ is lower-semi-continuous on X.

Let us prove now the second part of our theorem. Let ε be a positive number. Since φ is lower-semi-continuous on X, φ is a fortiori μ-measurable, so there exists a compact set $X_0 \subset X$ such that $\mu(X \setminus X_0) \leq \varepsilon$ and φ is continuous on X_0. If β is a minorant of f, then we have

$$\bigcup_{t \in X_0} \Phi(t) \subset K_0$$

where $K_0 = \{x \in G \mid g(x) \leq \rho - \beta\}$

$$\rho = \max_{t \in X_0} \varphi(t)$$

so that, for every t in X_0,

$$\Phi(t) = \{x \in K_0 \mid f(t,x) + g(x) \leq \varphi(t)\}$$

Therefore, we can conclude that Φ is upper-semi-continuous on X_0 because the graph of the restriction of Φ to X_0 is compact in the compact space $X_0 \times K_0$.

12 - <u>Corollary VII-12</u> - <u>With the same notation and hypotheses as in lemma VII-11 let</u> H <u>be a Banach space and let</u> G <u>be the vector space</u> H <u>equipped</u>

with the weak topology $\sigma(H,H')$. If, for every fixed t in T, the function $f_t : u \mapsto f(t,u)$ is convex on H and if g is strictly convex, then there exists a μ-measurable mapping s from T into H such that $s(t) \in \text{dom } f_t$ for all t in T.

Proof. Let z_0 be a fixed element in H. Let

$$\varphi(t,z_0) = \inf\{f(t,u) + g(z_0-u) \mid u \in G\}$$

$$\text{prox } (z_0 \mid f_t) = \{u \in G \mid f(t,u) + g(z_0-u) \leq \varphi(t,z_0)\}$$

Since g is strictly convex, the convex set prox $(z_0 \mid f_t)$ is reduced to a point which belongs to dom f_t for every fixed t in T. By the lemma VII-11 it suffices to prove that the mapping $t \mapsto \text{prox}(z_0 \mid f_t)$ from T to H is μ-measurable. But, this assertion is an easy consequence of the Dunford-Pettis-Phillips theorem ([7], p.304) which shows that a mapping from T to a Banach space is μ-measurable if and only if it is weakly μ-measurable (i.e. μ-measurable with respect to the weak topology).

13 - Corollary VII-13 - Let H be a reflexive Banach space and let G be the vector space H equipped with the weak topology $\sigma(H,H')$. If Γ is a pseudo-upper-semi-continuous multifunction from T with non empty convex closed values in G, then Γ admits a μ-measurable selection σ, that is, σ is a μ-measurable mapping from T to H such that $\sigma(t) \in \Gamma(t)$ for all t in T.

Proof. By virtue of a result proven by Lindentrauss ([10]), there exists on H a norm $\|.\|_1$ which is equivalent to the initial norm $\|.\|$ and this norm is strictly convex. Now, it suffices to apply corollary VII-12 by taking

$$f(t,x) = \begin{cases} 0 & \text{if} \quad x \in \Gamma(t) \\ +\infty & \text{if} \quad x \notin \Gamma(t) \end{cases}$$

$$g(x) = \|x\|_1^2$$

Remark. The result given in the corollary VII-13 was stated also in ([8]) under a stronger hypothesis (see for instance examples 2 and 3 following the definition of the pseudo-upper-semi-continuous property.)

14 - The results of measurability given above provide the duality theorem for decomposable vector spaces of measurable functions with values in a reflexive Banach space that follows. Let $\mathcal{L}_E$ (resp. $\mathcal{L}_{E'}$) be a vector space of μ-measurable functions from T to E (resp. E'_b) such that for every u (resp. v) belonging to $\mathcal{L}_E$ (resp. $\mathcal{L}_{E'}$) the scalar function $t \mapsto < u(t), v(t) >$ is μ-integrable, (remark that this function is μ-measurable thanks to our definition of measurability). The space $\mathcal{L}_E$ (or similarly $\mathcal{L}_{E'_b}$) is said to be decomposable if whenever u belongs to $\mathcal{L}_E$ and w belongs to $\mathcal{L}^\infty_E$ and A is a μ-measurable set in T, the function $\chi_A f + \chi_{T \setminus A}\, u$ belongs to $\mathcal{L}_E$.

Theorem VII-14 - Let f be a lower-semi-continuous function from $T \times E_\sigma$ to $]-\infty,+\infty]$. Suppose that f_t is convex, $\mathcal{L}_E$ is decomposable and the integral functional on $\mathcal{L}_E$,

$$I_f(u) = \begin{cases} \int_T f(t,u(t))\mu(dt) & \text{if } \int_T f(t,u(t))^+\mu(dt) < +\infty \\ +\infty & \text{if } \int_T f(t,u(t))^+\mu(dt) = +\infty \end{cases}$$

is finite for at least one u_0 in $\mathcal{L}_E$.

Then the convex integral functional I_{f^*} defined on $\mathcal{L}_{E'_b}$ by

$$I_{f^*}(v) = \begin{cases} \int_T f^*(t,v(t))\,\mu(dt) & \text{if } \int_T f^*(t,v(t))^+\mu(dt) < +\infty \\ +\infty & \text{if } \int_T f^*(t,v(t))^+\mu(dt) = +\infty \end{cases}$$

is conjugate to the integral functional I_f on $\mathcal{L}_{E'}$ that is

$$I_{f^*}(v) = \sup\,\{< v,u > - I_f(u) \mid u \in \mathcal{L}_E\}$$

for every v in $\mathcal{L}_{E'_b}$.

Proof. Thanks to the measurability result established above, the arguments given in the proof of the theorem VII-7 can be used here. For the convenience of the reader we shall sketch a short proof. First, for every v in $\mathcal{L}_{E'_b}$, $t \mapsto f^*(t,v(t))$ is μ-measurable on T by lemma VII-9. We have to prove that

$$\int_T f^*(t,v(t))\mu(dt) \leq \sup\{< u,v > - I_f(u) \mid u \in \mathcal{L}_E\}$$

for every fixed v in $\mathcal{L}_{E'_b}$. Let β be a real number such that $\beta < I_{f^*}(v)$. We prove the existence of a function u in $\mathcal{L}_E$ such that

$$< u,v > - I_f(u) \geq \beta$$

By the construction given in the proof of theorem VII-7, there exists a real integrable function α_o such that

$$< v(t), u_o(t) > - f(t,u_o(t)) \geq \alpha_o(t)$$

for all $t \in T$ and a real integrable function γ such that

$$\begin{cases} \int_T \gamma(t)\mu(dt) > \beta \\ \gamma(t) < f^*(t,v(t)) \quad \text{for all} \quad t \in T. \end{cases}$$

Define the multifunction Γ from T to E by

$$\Gamma(t) = \{x \in E \mid < v(t),x > - f(t,x) \geq \gamma(t)\}$$

Since v and γ are μ-measurable, we can find an increasing sequence (T_n) of compact sets such that $\mu(T \setminus \bigcup_n T_n) = 0$ and the restriction of v (resp. γ) to each T_n is continuous. Therefore, the restriction of Γ to each T_n is a multifunction with non empty convex closed values in E_σ such that the graph $G(\Gamma_{|T_n})$ of $\Gamma_{|T_n}$ is sequentially

$$\{(t,x) \in T_n \times E_\sigma \mid < v(t),x > - f(t,x) \geq \gamma(t)\}$$

closed in the product space $T_n \times E_\sigma$. The corollary VII-13 of lemma VII-12 then implies the existence of a μ-measurable mapping s_n from T_n to E such that $s_n(t) \in \Gamma(t)$ for all $t \in T_n$. For each n let

$$u_n(t) = \begin{cases} s_n(t) & \text{if} \quad t \in T_n \\ u_o(t) & \text{if} \quad t \in T \setminus T_n \end{cases}$$

Then u_n belongs to $\mathcal{L}_E$ by the decomposability hypothesis. For each n we have

$$< v(t),\ u_n(t) > - f(t,u_n(t)) \geq \gamma(t) \quad \text{if} \quad t \in T_n$$

$$< v(t),\ u_n(t) > - f(t,u_n(t)) \geq \alpha(t) \quad \text{if} \quad t \in T \setminus T_n$$

Thus

$$< v,u_n > - I_f(u_n) \geq \int_{T_n} \gamma(t)\mu(dt) + \int_{T \setminus T_n} \alpha(t)\mu(dt) \geq \beta$$

when n is large ; so we can conclude that

$$I_{f*}(v) = \sup\{< v,u > - I_f(u) \mid u \in \mathcal{L}_E\}.$$

<u>Notation</u>. Henceforth, we denote by L_E (resp. $L_{E'_b}$) the quotient of $\mathcal{L}_E$ (resp. $\mathcal{L}_{E'_b}$) by the equivalence relation "equality almost everywhere".

<u>Remark</u>. In the case of a non separable reflexive Banach space E, the conjugate function f* of f defined on $T \times E'$

$$f^*(t,y) = \sup\{< x,y > - f(t,x) \mid x \in E\}$$

is not lower-semi-continuous on $T \times E'_s$, hence we cannot expect to have the same situation as in the theorem VII-7. However we give below an example encountered in a new kind an evolution equation ([15], [16], [17], [19]) in which the whole of the duality theorem VII-7 holds.

<u>Example</u>. Let T be the interval $[0,1]$ and μ be the Lebesgue measure on T. A multifunction C from T with non empty convex closed values in E is of <u>continuous variation</u> on T if there exists a real continuous function r, defined on T such that

$$h(C(t),\ C(\tau)) \leq |r(t) - r(\tau)|$$

for every $t \in T$ and $\tau \in T$ where h is the Hausdorff distance on the non empty convex closed sets in E. If C is of continuous variation, the then the domain D of the support functional $x' \mapsto \delta^*(x',\ C(t))$ of $C(t)$ does not depend on t and it is easy to express the Hausdorff distance in terms of support functionals, namely

$$h(C(t), C(\tau)) = \sup_{x' \in B' \cap D} |\delta^*(x', C(t)) - \delta^*(x', C(\tau))|$$

where B' denotes the unit ball in E'.

Moreover, the graph of C,

$$\{(t,x) \in T \times E \mid x \in C(t)\}$$

is closed in $T \times E_\sigma$ where E_σ is the vector space E endowed with the weak topology $\sigma(E,E')$, since we have

$$C(t) = \bigcap_{x' \in D} C_{x'}(t)$$

where

$$C_{x'}(t) = \{x \in E \mid < x',x > \leq \delta^*(x', C(t))\}$$

The graph of each multifunction $C_{x'}$ $(x' \in D)$ is closed in $T \times E_\sigma$, so the graph of C is closed too. In addition, C is lower-semi-continuous on T. For every fixed x and y in E and every fixed τ and t in $[0,1]$, we have

$$|d(x,C(t)) - d(y, C(\tau))| \leq \sup_{x' \in B \cap D} | < x',x > - \delta^*(x',C(t)) -$$

$$- < x',y > + \delta^*(x',C(t)) | \leq \|x-y\| + |r(t) - r(\tau)|$$

This implies in particular that for every fixed x in E, the function $t \mapsto d(x, C(t))$ is continuous on T so that C is lower-semi-continuous on T. Now, we take

$$f(t,x) = \begin{cases} 0 & \text{if } x \in C(t) \\ +\infty & \text{if } x \notin C(t) \end{cases}$$

$$f^*(t,y) = \delta^*(y, C(t))$$

Then f is lower-semi-continuous on $T \times E_\sigma$ since the graph of C is closed in $T \times E_\sigma$. Let

$$S_C = \{u \in L^1_E(T,\mu) \mid u(t) \in C(t) \quad \text{a.e}\}$$

Since C is lower-semi-continuous, there exists by Michael's theorem ([11]) a continuous selection for C, that is a continuous mapping s

from T to E such that $s(t) \in C(t)$ for all $t \in T$. So S_C is a closed convex non empty subset of $L^1_E(T,\mu)$. Clearly, the integral functional

$$I_f(u) = \int_T f(t,u(t))dt, \quad u \in L^1_E(T,\mu)$$

is the indicator function of S_C,

$$I_f(u) = \begin{cases} 0 & \text{if } u \in S_C \\ +\infty & \text{if } u \notin S_C \end{cases}$$

and theorem VII-14 shows that the integral functional

$$I_{f^*}(v) = \int_T f^*(t,v(t))dt, \quad v \in L^\infty_{E'_b}(T,\mu)$$

is the support function of the convex closed set S_C. So, the above example allows us to formulate now the following corollary of theorem VII-14.

15 - <u>Corollary VII-15</u> - <u>Suppose that</u> L_E (<u>resp.</u> $L_{E'_b}$) <u>is</u> $L^p_E(T,\mu)$ (<u>resp.</u> $L^q_{E'_b}(T,\mu)$) <u>with</u> $1 \leq p < +\infty$, $\frac{1}{p} + \frac{1}{q} = 1$, <u>neither the convex integral functional</u> I_f <u>on</u> $L^p_E(T,\mu)$ <u>nor</u> I_{f^*} <u>on</u> $L^q_{E'_b}(T,\mu)$ <u>is identically</u> $+\infty$. <u>Then these convex integral functionals are conjugate to each other, that is</u>

$$I_{f^*}(v) = \sup\{< u,v > - I_f(u) \mid u \in L^p_E(T,\mu)\}$$

<u>for every</u> v <u>in</u> $L^q_{E'_b}(T,\mu)$, <u>and</u>

$$I_f(u) = \sup\{< u,v > - I_{f^*}(v) \mid v \in L^q_{E'_b}(T,\mu)\}$$

<u>for every</u> u <u>in</u> $L^p_E(T,\mu)$.

<u>Proof</u>. It suffices to prove that I_f is lower-semi-continuous on L^p_E. Let λ be a real number, we must check that the set

$$\{u \in L^p_E \mid I_f(u) \leq \lambda\}$$

is closed in L^p_E. Let (u_n) be a sequence in L^p_E such that $I_f(u_n) \leq \lambda$ for every n. By hypothesis, there exists $v_0 \in L^q_E$ such that $\int_T f^*(t, v_0(t))d\mu(t) < +\infty$. Assume that (u_n) converges to u_0 in L^p_E. By extracting a subsequence of (u_n), one can suppose that (u_n) converges a.e. to u_o so that, by applying Fatou's lemma to the sequence

$$t \mapsto f(t, u_n(t)) - < u_n(t), v_o(t) > + f^*(t, v_o(t))$$

we obtain

$$\int_T \liminf_{n \to \infty} [f(t,u_n(t)) - < u_n(t), v_o(t) > + f^*(t,v_o(t))]\mu(dt)$$
$$\leq \lambda - \int_T [< u_o(t),v_o(t) > - f^*(t,v_o(t))]\mu(dt)$$

So we have

$$\int_T f(t,u_o(t))\mu(dt) \leq \lambda$$

and the proof is complete.

§ 4 - APPLICATIONS OF THE DUALITY THEOREM OF CONVEX INTEGRAL FUNCTIONALS.

In this section we give some applications of the results obtained in § 2. First we prove that Strassen's theorem ([23] theor. 1, p.424) is a consequence of theorem VII-7 and theorem I-29.

16 - Application of theorem VII-7 to Strassen's theorem.

We obtain Strassen's theorem easily by taking a complete probability space $(\Omega, \mathcal{A}, \mu)$, a separable Banach space E, and a normal convex integrand f defined on $\Omega \times E$ with real values such that $x \mapsto f(\omega,x)$ is continuous positively homogeneous on E for every fixed ω in Ω. In addition, we suppose that there exists a positive number k such that $|f(\omega,x)| \leq k\|x\|$ for all ω in Ω and all x in E. Now, let A be the natural continuous injective mapping from E to $L^1_E(\Omega, \mathcal{A}, \mu)$. Then the adjoint A^* of A is defined by

$$< v,Ax > = \int_\Omega < v(\omega),x > \mu(d\omega) = < x, \int_\Omega v(\omega)\mu(d\omega) >$$
$$= < x, A^*(v) >$$

for all $v \in L^\infty_{E'_s}(\Omega, \mathcal{A}, \mu)$. So A^* is the mapping $v \mapsto \int_\Omega v(\omega)\mu(d\omega)$ from $L^\infty_{E'_s}(\Omega, \mathcal{A}, \mu)$ to E'_s. Since the convex integral functional

$$I_f(u) = \int_\Omega f(\omega,u(\omega))\mu(d\omega)$$

is clearly continuous on $L^1_E(\Omega,\mathcal{A},\mu)$ and I_f and I_{f^*} are conjugate to each other by theorem VII-7, we deduce from theorem I-29, that, for every x in E, the subdifferential $\partial(I_f \circ A)(x)$ of $I_f \circ A$ at the point x, is equal to the set

$$\{A^* v \mid v \in L^\infty_{E'_s}(\Omega,\mathcal{A},\mu)\ ,\ v(\omega) \in \partial f_\omega[(Ax)(\omega)] \text{ a.e.}\}$$

In particular by taking $x = 0$, one has

$$\partial(I_f \circ A)(0) = \{x' \in E' \mid \langle x',x\rangle \leq \int_\Omega f(\omega,x)\mu(d\omega),\ \forall x \in E\}$$

$$= \{\int_\Omega v(\omega)\mu(d\omega) \mid v(\omega) \in \partial f_\omega(0) \text{ a.e.}\}$$

17 - Application of theorem VII-7 to a result of disintegration of measures.

We particularize below assumptions given in theorem VII-7 and theorem I-29.

U : compact metrizable space

(T,μ) : compact space equipped with Radon positive measure μ.

E : the separable Banach space $\mathcal{C}(U)$ of real continuous functions on U endowed with the topology of uniform convergence.

Z : compact space,

G : the Banach space $\mathcal{C}(T \times Z \times U)$ of real continuous functions on $T \times Z \times U$ equipped with the topology of uniform convergence,

v : μ-measurable mapping from T to Z,

Γ : μ-measurable multifunction from T with non empty compact values in U,

Σ : μ-measurable multifunction from T with non empty convex compact values in the compact space of Radon probability measure on U, $\mathcal{M}^1_+(U)$, equipped with the vague topology,

$$\Sigma(t) = \{\nu \in \mathcal{M}^1_+(U) \mid \nu(\Gamma(t)) = 1\}\ ,\ \forall t \in T,$$

f_t : support function of $\Sigma(t)$:

$$f(t,x) = \sup_{\nu\in\Sigma(t)} \langle x,\nu\rangle = \sup_{u\in\Gamma(t)} x(u),\ \forall t \in T,\ \ \forall x \in \mathcal{C}(U),$$

A : the mapping from $\mathcal{C}(T \times Z \times U)$ to $L^1_{\mathcal{C}(U)}(T,\mu)$ defined as **follows.** **With each** $W \in \mathcal{C}(T \times Z \times U)$ let us associate the element $A(W)$ in $L^1_{\mathcal{C}(U)}(T,\mu)$ by formula

$$A(W)(t) = W(t,v(t),.)$$

Note that the adjoint mapping A^* of A is given by

$$A^*(\lambda) = \int_T (\epsilon_t \otimes \epsilon_{v(t)} \otimes \lambda_t) d\mu(t) \ , \ \lambda \in L^\infty_{\mathcal{M}(U)}(T,\mu)$$

where ϵ_a denotes the Dirac mass at the point a and $\mathcal{M}(U)$ the weak dual of $\mathcal{C}(U)$; for every $W \in \mathcal{C}(T \times Z \times U)$, we have

$$< W, A^*(\lambda) > = < A(W), \lambda > = \int_T [\int_U W(t,v(t),u) d\lambda_t(u)] d\mu(t)$$
$$= \int_T < W, \epsilon_t \otimes \epsilon_{v(t)} \otimes \lambda_t > d\mu(t)$$
$$= < W, \int_T (\epsilon_t \otimes \epsilon_{v(t)} \otimes \lambda_t) \, d\mu(t) >$$

But we have

$$\sup \{W(t,z,u) \mid (t,z,u) \in t \times v(t) \times \Gamma(t)\}$$
$$= \sup \{W(t,v(t),u) \mid u \in \Gamma(t)\}$$
$$= f(t, A(W)(t))$$

Then, by using again theorem VII-7 and theorem I-29, we obtain the equality,

$$\partial (I_f \, oA)(0) = \{\nu \in \mathcal{M}(T \times Z \times U) \mid < \nu, W > \leq \int_T f(t,A(W)(t)) \, d\mu(t), \ \forall W \in \mathcal{C}(T \times Z \times U)\}$$
$$= \{\int_T (\epsilon_t \otimes \epsilon_{v(t)} \otimes \lambda_t) \, d\mu(t) \mid \lambda_t \in \partial f_t(0) = \Sigma(t) \text{ a.e.}\}$$

Thanks to the preceding formula we can state now a result of desintegration of measures (Compare with ([12], [24] [26]).

Let T and Z be two compact spaces, and U a compact metrizable space, v a universally measurable mapping from T to Z, Γ a universally measurable multifunction from T with non empty compact values in U. Let G be the graph of the multifunction $v \times \Gamma$ and χ_G the characteristic function of the universally measurable set G. Let ν be a Radon *positive measure on* $T \times Z \times U$ *and* p *the projection of* $T \times Z \times U$ *onto* T. *Put*

$$\mu = \int_{T \times Z \times U} \varepsilon_{p(t,z,u)} \chi_G(t,z,u) d\nu(t,z,u)$$

Then, there exists a scalarly μ*-measurable mapping* λ *from* T *to the set* $\mathcal{M}_+^1(U)$ *of probability Radon measures on* U *such that* $\lambda_t(\Gamma(t)) = 1$ *for all* $t \in T$ *and such that the measure* $\chi_G \nu$ *admits the integral representation*

$$\chi_G \nu = \int_T \varepsilon_t \otimes \varepsilon_{v(t)} \otimes \lambda_t \quad d\mu(t).$$

Proof. Let π be the restriction of p to G.

For each $W \in \mathcal{C}(T \times Z \times U)$, let us consider the function $\hat{W}$ defined on T by

$$\hat{W}(t) = \sup\{W(t,z,u) | (t,z,u) \in \pi^{-1}(t)\}$$

$$= \sup_{u \in \Gamma(t)} W(t, v(t), u)$$

Since Γ is universally measurable, $\hat{W}$ is universally measurable too by noting that there exists a sequence $(u_n)_{u \in \mathbb{N}}$ of universally measurable selections of Γ such that the set $\{u_n(t) | n \in \mathbb{N}\}$ is dense in $\Gamma(t)$ for each $t \in T$. By virtue of a Bourbaki theorem ([2] , theor. 2, p.40) we have

$$\int_{T \times Z \times U} \hat{W}(p(t,z,u)) \chi_G(t,z,u) d\nu(t,z,u) = \int_T \hat{W}(t) d\mu(t)$$

for all $W \in \mathcal{C}(T \times E \times U)$. But

$$W \chi_G \leq \hat{W} o\, p \;\; \chi_G \quad \text{for all} \quad W \in \mathcal{C}(T \times Z \times U)$$

Hence

$$\int_{T \times Z \times U} W \chi_G \, d\nu \leq \int_T \hat{W} \, d\mu$$

So $\chi_G \nu$ belongs to $\partial (I_f \, oA)(0)$ by using the above notations. Therefore we conclude that

$$\chi_G \nu = \int_T (\varepsilon_t \otimes \varepsilon_{v(t)} \otimes \lambda_t) \, d\mu(t)$$

with $t \mapsto \lambda_t$ scalarly μ-measurable such that $\lambda_t(\Gamma(t)) = 1$ for all $t \in T$.

18 - Applications of the duality theorem to the existence of a solution of an evolution equation.

We first state the existence of a solution of a stochastic evolution equation $-X'(\omega,t) \in \partial\varphi^*(\omega,t,X(\omega,t))$ arose in sectionn 14 of chapter VI.

Theorem VII-18 - With the hypotheses and notation of theorem VI-14 assume that E is real separable Hilbert space and denote by $\varphi(\omega,t,.)$ and $\varphi^*(\omega,t,.)$ the support function and the indicator function of $\Gamma(\omega,t)$

$$\varphi(\omega,t,x) = \sup\{<x,u> \mid u \in \Gamma(\omega,t)\}$$

$$\varphi^*(\omega,t,v) = \begin{cases} 0 & \text{if } v \in \Gamma(\omega,t) \\ +\infty & \text{if } v \notin \Gamma(\omega,t) \end{cases}$$

and let $\partial\varphi^*(\omega,t,v)$ be the subdifferential of the convex function $\varphi^*(\omega,t,v)$ at v,

$$\partial\varphi^*(\omega,t,v) = \{z \in E \mid \varphi(\omega,t,z) + \varphi^*(\omega,t,v) \leq <z,v>\}$$

Then there exists a mapping X from $\Omega \times [0,T]$ to E with the following conditions :

(i) $X(\omega,t) = y(\omega) + \int_0^t X'(\omega,s)ds$

for all (ω,t), where X' is a $P \otimes ds$-integrable mapping from $\Omega \times [0,T]$ to E

(ii) There exists a $P \otimes ds$ null set, M, such that

$$0 \in X'(\omega,t) + \partial\varphi^*(\omega,t,X(\omega,t))$$

for all $(\omega,t) \notin M$.

Proof. We apply here the notation and the arguments in the proof off theorem VI-14. By the construction given in the proof of this theorem there exists a sequence (X_n) of mappings from $\Omega \times [0,T]$ to E and a sequence of step functions (θ_n) from $[0,T]$ to $[0,T]$ with the following properties:

$$\begin{cases} X_n(\omega,t) = y_0(\omega) + \int_0^t X'_n(\omega,s)ds, \ \|X'_n(\omega,s)\| \le g(s) \text{ a.e} \\ -X'_n(\omega,t) \in \partial\varphi^*(\omega, \theta_n(t), X_n(\omega,\theta_n(t))) \\ \lim_{n\to\infty} \theta_n(t) = t \end{cases}$$

for all $(\omega,t) \in \Omega \times [0,T]$. Indeed, let S_n be a subdivision of $[0,T]$ which is used in the proof of theorem VI-14 ~~i~~-14,

$$0 = t_0^n <\ldots< t_i^n \ldots<\ldots< t_{k_n}^n = T$$

For every t in $[0,T]$, denote by $j_n(t)$ the value of $i(1 \le i \le k_n)$ such that $t \in]t_{i-1}^n, t_i^n]$; for every $(\omega,t) \in \Omega \times [0,T]$, let

$$\begin{cases} \theta_n(t) = t_{j_n(t)}^n \\ X_n(\omega, \theta_n(t)) = y_{j_n(t)}^n(\omega) \end{cases}$$

Then we have

$$y_{j_n(t)}^n(\omega) = \text{proj}(y_{j_n(t)-1}^n(\omega) \,|\, \Gamma(\omega,\theta_n(t))$$

According to the basic property of Moreau's proximation mapping ([13] , prop.4.a., p.280),

$$\frac{y_{\theta_n(t)}^n(\omega) - y_{j_n(t)-1}^n(\omega)}{r(\theta_n(t)) - r(t_{j_n(t)-1}^n)} \in \partial\varphi^*(\omega, \theta_n(t), X_n(\omega,\theta_n(t))$$

since the second member is a cone. By the proof of theorem VI-14, we can extract from the sequence (X_n) a subsequence, still denoted by (X_n), which converges in the compact metrizable set

$$\mathcal{X} = \{y \in \mathcal{C}([0,T], L_E^\infty(\Omega,\mathcal{A},P)) \,|\, Y(\omega,t) = y_0(\omega) + \int_0^t Y'(\omega,x)ds, \ Y' \in \mathcal{S}\}$$

$$\text{where } \mathcal{S} = \{Y' \in L_E^1(\Omega\times[0,T] \,|\, \|Y'(\omega,s)\| \le g(s) \text{a.e.}\}$$

to an element $X \in \mathcal{X}$. This implies that

$$X(\omega,t) = Y_0(\omega) + \int_0^t X'(\omega,s)\, ds, \ \|X'(\omega,s)\| \le g(s) \quad \text{a.e.}$$

But the sequence (X'_n) is relatively compact with respect to the weak topology of $L_E^1(\Omega \times [0,T], \mathcal{A} \hat{\otimes} \mathcal{C}, P \otimes ds)$ by corollary V-4. So, by virtue of Eberlein-Smulian's theorem, we can extract from (X'_n) a subsequence which converges weakly to Y' with $\|Y'(\omega,s)\| \leq g(s)$ a.e., since the set

$$\mathcal{S} = \{Y' \in L^1_E(\Omega \times [0,T]) \mid \|Y'(\omega,s)\| \leq g(s) \text{ a.e.}\}$$

is compact with respect to this weak topology ; and Y' must coincide necessarily with X'. We now prove that X satisfies the required properties of our theorem. Clearly, it remains to prove (ii).

For every $U \in L^1_E(\Omega \times [0,T])$ and $V \in L^\infty_E(\Omega \times [0,T])$, put

$$I_\varphi(U) = \int_{\Omega\times[0,T]} \varphi(\omega,t,U(\omega,t)) \; P(d\omega) \otimes dt$$

$$I_{\varphi^*}(V) = \int_{\Omega\times[0,T]} \varphi^*(\omega,t,V(\omega,t)) \; P(d\omega) \otimes dt$$

and, if $U \in \mathcal{S}$, put

$$q(U) = \int_{\Omega\times[0,T]} < U(\omega,t),\, y_0(\omega) + \int_0^t U(\omega,s)\, ds > P(d\omega) \otimes dt$$

Then, φ and φ^* are convex normal integrands on $\Omega\times[0,T] \times E$. For this, it suffices to verify that the multifunction Γ is $\mathcal{A} \otimes \mathcal{C}$ measurable.

For every fixed $x \in E$, the function

$$(\omega,t) \to d(x, \Gamma(\omega,t)) = \inf_{u\in\Gamma(\omega,t)} \|x-u\|$$

is $\mathcal{A}$-measurable on Ω for every fixed t in $[0,T]$ and is continuous on $[0,T]$ for every fixed ω in Ω because we have by a very simple calculation (see the example following theorem VII-14)

$$|d(x,\Gamma(\omega,t)) - d(x, \Gamma(\omega,\tau))| \leq |r(t) - r(\tau)|$$

for all (τ,t) in $[0,T]^2$. So, Γ is $\mathcal{A} \otimes \mathcal{C}$-measurable by theorem III-9. As Γ admits a $\mathcal{A}\otimes\mathcal{C}$-measurable and bounded selection by theorem VI-14, hence, according to the duality theorem VII-7, we conclude that the integral convex functionals I_φ and I_{φ^*} are proper convex conjugate to each other. Moreover, for every $U \in \mathcal{S}$, we have, by applying Fubini's theorem,

$$\int_{\Omega\times[0,T]} < U(\omega,t), \int_0^t U(\omega,s)ds > P(d\omega) \otimes dt = \frac{1}{2} \int_\Omega \left\|\int_0^T U(\omega,t)dt\right\|^2 P(d\omega)$$

Now we assert that the convex function

$$\psi : U \mapsto \int_{\Omega} \left\|\int_0^T U(\omega,t)dt\right\|^2 P(d\omega) \qquad (U \in \mathcal{S})$$

is semi-continuous on the convex closed set $\mathcal{S}$ of $L^1_E(\Omega \times [0,T], \mathcal{A} \hat{\otimes} \mathcal{T}, P \otimes dt)$ for the norm topology.

As we have already observed, the set $\mathcal{S}$ is equal to the set of mappings $f : \Omega \times T \to E$ defined by,

$$f(\omega,t) = \pi_{\Omega}(\omega)g(t)\ V(\omega,t)$$

where π_{Ω} is the characteristic function of Ω and V is a $\mathcal{A} \hat{\otimes} \mathcal{T}$-measurable mapping from $\Omega \times [0,T]$ to E such that $\|V(\omega,t)\| \leq 1$, $P \otimes dt$ - a.e.

Let $\nu = g\ ds$. For every element V in $L^2_E(\Omega \times [0,T], \mathcal{A} \hat{\otimes} \mathcal{T}, P \otimes \nu)$, one has by the well known Jensen's inequality,

$$\left\|\int_0^T V(\omega,s) \frac{\nu(ds)}{r(T)}\right\|^2 \leq \int_0^T \|V(\omega,s)\|^2 \frac{\nu(ds)}{r(T)}$$

Hence we have

$$\left\|\int_0^T (V(\omega,s)\nu(ds)\right\|^2 \leq \left(\int_0^T \|V(\omega,s)\|^2\ \nu(ds)\right) \int_0^T 1\ \nu(ds)$$

By integrating we obtain

$$\int_{\Omega} \left(\left\|\int_0^T V(\omega,s)\nu(ds)\right\|^2 P(d\omega)\right) \leq \left(\int_0^T \nu(ds)\right) \int_{\Omega}\left(\int_0^T \|V(\omega,s)\|^2\ \nu(ds)\right)P(d\omega)$$

$$= \left(\int_0^T \nu(ds)\right)N_2(V)^2$$

where $N_2(V)$ is the norm in $L^2_E(\Omega \times [0,T], \mathcal{A} \hat{\otimes} \mathcal{T}, P \otimes \nu)$.

This implies that the convex function

$$J : V \mapsto \int_{\Omega} \left(\left\|\int_0^T V(\omega,s)\nu(ds)\right\|^2\right)P(d\omega)$$

is continuous on the space $L^2_E(\Omega \times [0,T], \mathcal{A} \hat{\otimes} \mathcal{T}, P \otimes \nu)$ with respect to the norm topology. This enables us to prove the lower semi-continuity of the convex function ψ on the closed convex set $\mathcal{S}$ of

$L^1_E(\Omega \times [0,T], \mathcal{A} \otimes \mathcal{T}, P \otimes ds)$. Let $r \in \mathbb{R}$. We have to check that the set

$$\{U \in \mathcal{S} \mid \psi(U) \leq r\}$$

is closed. Let (U_n) be a sequence in $\mathcal{S}$ such that $\psi(U_n) \leq r$ and suppose that (U_n) converges to an element $\tilde{U}$ in $\mathcal{S}$. There exists a sequence (V_n) of $\mathcal{A} \hat{\otimes} \mathcal{T}$-measurable mappings from $\Omega \times [0,T]$ to E such that

$$\begin{cases} \|V_n(\omega,t)\| \leq 1 \quad , P \otimes ds \quad a.e. \\ U_n(\omega,t) = 1_\Omega(\omega)\ g(t)\ V_n(\omega,t) \ , P \otimes ds \quad a.e. \end{cases}$$

Moreover we may suppose, by extracting a subsequence, that

$$\lim_{n \to \infty} U_n(\omega,t) = \tilde{U}(\omega,t) \quad , \quad P \otimes ds - a.e.$$

Let $\tilde{V}$ be a $\mathcal{A} \hat{\otimes} \mathcal{T}$-measurable mapping from $\Omega \times [0,T]$ to E such that

$$\begin{cases} \|\tilde{V}(\omega,t)\| \leq 1 \quad , \quad P \otimes ds - a.e. \\ \tilde{U}(\omega,t) = 1_\Omega(\omega)\ g(t)\ \tilde{V}(\omega,t), \ P \otimes ds \quad a.e. \end{cases}$$

So we have $\lim_{n \to \infty} V_n(\omega,t) = \tilde{V}(\omega,t)$, $P \otimes ds$ - a.e, this implies a fortiori that

$$\lim_{n \to \infty} V_n(\omega,t) = \tilde{V}(\omega,t), \quad P \otimes \nu - a.e.$$

We deduce that the sequence (V_n) convergesto $\tilde{V}$ with respect to the topology $\sigma(L^\infty_E(P \otimes \nu), L^1_E(P \otimes \nu))$ and then (V_n) converges to $\tilde{V}$ with respect to the topology $\sigma(L^\infty_E(P \otimes \nu), L^2_E(P \otimes \nu))$. Let $\mathcal{V}$ be the unit ball of $L^\infty_E(P \otimes \nu)$. Then we have

$$\{U \in \mathcal{S} \mid \psi(U) \leq r\} = \{V \in \mathcal{V} \mid J(V) \leq r\}$$

Since $\mathcal{V}$ is convex and closed in $L^2_E(P \otimes \nu)$ and since J is convex and continuous on the space $L^2_E(P \otimes \nu)$, we conclude that $\psi(\tilde{U}) = J(\tilde{V}) \leq r$. It follows that the convex function q defined on $\mathcal{S}$

$$q(U) = \int_{\Omega \times [0,T]} < U(\omega,t),\ y_0(\omega) + \int_0^t U(\omega,s)ds > P(d\omega) \otimes dt$$

is lower-semi-continuous on the convex closed set $\mathcal{S}$. So, for every n, we have

$$I_\varphi(-X_n') + \langle X_n', X_n \rangle = I_\varphi(-X_n') + q(X_n')$$

By using the relation $-X_n'(\omega,t) \in \partial \varphi^*(\omega, \theta_n(t), X_n(\omega, \theta_n(t))$, we have

$$(1) \quad \begin{cases} X_n(\omega,\theta_n(t)) \in \Gamma(\omega, \theta_n(t)) \\ \varphi(\omega,\theta_n(t), -X_n'(\omega,t)) + \langle X_n'(\omega,t), X_n(\omega,\theta_n(t))\rangle = 0 \end{cases}$$

By a simple calculation we have following inequalities

$$(2) \quad \left\| \int_{\Omega\times[0,t]} \langle X_n'(\omega,t), \int_t^{\theta_n(t)} X_n'(\omega,s)ds \rangle P(d\omega) \otimes dt \right\| \le \int_0^T g(t)|r(\theta_n(t)) - r(t)|dt$$

$$(3) \quad \int_{\Omega\times[0,T]} [\varphi(\omega,t, -X_n'(\omega,t)) - \varphi(\omega,\theta_n(t), -X_n'(\omega,t))]P(d\omega) \otimes dt \le \int_0^t |r(\theta_n(t)) - r(t)|g(t)dt$$

From relations (1), (2), (3), we deduce that

$$(4) \quad I_\varphi(-X_n') + \langle X_n', X_n \rangle = I_\varphi(-X_n') + q(X_n') \le 2\int_0^T |r(\theta_n(t)) - r(t)|g(t)dt$$

As I_φ and q are lower-semi-continuous with respect to the weak topology of $L_E^1(\Omega \times [0,T], \mathcal{A}\hat{\otimes}\mathcal{T}, P\otimes dt)$, it follows from (4), that

$$(5) \quad I_\varphi(-X') + \langle X,X' \rangle \le \liminf_{n\to\infty}(I_\varphi(-X_n') + \langle X_n', X_n \rangle) \le 0$$

From (5) and the duality theorem VII-7, we obtain

$-X'(\omega,t) \in \partial \varphi^*(\omega,t,X(\omega,t))$, $P\otimes dt$ - a.e. and the proof is complete.

19 - A variant of theorem VII-18.

When Γ depends only on t (this is the deterministic case), the assumptions of theorem VII-18 can be weakened ; namely we have the following variant of theorem VII-18.

Theorem VII-19 - Let Γ be a multifunction from $[0,T]$ with non empty closed convex values in a real Hilbert space E with the following property : for $0 \le \tau \le t \le T$, $h(\Gamma(t), \Gamma(\tau)) \le \int_\tau^t g(s)\,ds$ where g is a strictly ds-integrable function on $[0,T]$ and h is Hausdorff distance defined on the set of closed convex subsets of E.

Let y_0 *be a fixed point in* $\Gamma(0)$. *Then, there exists a mapping* X *from* $[0,T]$ *to* E *such that* :

(i) $$X(t) = y_0 + \int_0^t X'(s)\, ds \quad (t \in [0,T])$$

where X' *is a* ds-*integrable function from* $[0,T]$ *to* E

(ii) *There exists a* ds-*negligible set* N *such that*

$$- X'(t) \in \partial\, \varphi^*(t, X(t))$$

for all $t \notin N$; *where* $\partial\, \varphi^*(t,.)$ *is the subdifferential of the indicator function* $\varphi^*(t,.)$ *of* $\Gamma(t)$.

Proof. For the convenience of the reader we only sketch the proof since the arguments in the proof of theorem VI-14 and VII-18 are applicable here. By the construction given in theorem VI-14 there exists a sequence (X_n) of mappings from $[0,T]$ to E and a sequence (θ_n) of step functions from $[0,T]$ to $[0,T]$ such that

$$\begin{cases} X_n(t) = y_0 + \int_0^t X_n'(s)\, ds \ , \ \|X_n'(s)\| \leq g(s) \text{ a.e.} \\ X_n'(t) \in \partial\, \varphi^*(\theta_n(t), X_n(\theta_n(t))) \\ \lim_{n \to \infty} \theta_n(t) = t \end{cases}$$

for all $t \in [0,T]$. Since the sequence (X_n') is relatively compact with respect to the weak topology of $L^1_E[0,T]$ by theorem V-10, so, by virtue of Eberlein-Smulian's theorem, we can extract from (X_n') a subsequence with converges weakly to X' with $\|X'(s)\| \leq g(s)$ a.e. since the set

$$\mathcal{S} = \{Y' \in L^1_E\, [0,T] \mid \|Y'(s)\| \leq g(s) \text{ a.e.}\}$$

is weakly compact in $L^1_E[0,T]$ by theorem V-10.

Put $X(t) = y_0 + \int_0^t X'(s)\, ds \qquad t \in [0,T])$.

We shall show that X satisfies property (ii) there establishing our theorem. For every U in $L^1_E\,[0,T]$ and every V in $L^\infty_E[0,T]$, put

$$I_\varphi(U) = \int_0^T \varphi(t,U(t))dt$$

$$I_{\varphi}(V) = \int_0^T \varphi^*(t, V(t))\, dt$$

and, if $U \in \mathcal{S}$, put

$$q(U) = \int_0^T < U(t),\ y_0 + \int_0^t U(s)\ ds > dt = \int_0^T < U(t), y_0 > dt + \frac{1}{2}\left\|\int_0^T U(t) dt\right\|^2$$

Since the multifunction Γ is of continucus variation, we know by the example which follows theorem 14 that the integral convex functionals

$$I_\varphi(U) = \int_0^T \varphi(t,U(t))dt,\ U \in L^1_E\ [0,T]$$

$$I_{\varphi^*}(V) = \int_0^T \varphi^*(t,V(t))dt,\ V \in L^\infty_E[0,T]$$

are proper convex functions conjugate to each other. Moreover the convex function q is lower-semi-continuous on the weakly convex compact set $\mathcal{S}$ of $L^1_E[0,T]$. For every n, we have

$$I_\varphi(-X'_n) + < X'_n, X_n > = I_\varphi(-X'_n) + q(X'_n)$$

$$\leq 2 \int_0^T |r(\theta_n(t)) - r(t)|\ g(t)\ dt.$$

We deduce easily from the lower-semi-continuity of I_φ and q that

$$I_\varphi(-X') + < X',X > \leq \liminf_{n \to \infty} [I_\varphi(-X'_n) + < X'_n,X_n >] \leq 0$$

This inequality and the corollary VII-15 show that

$$-X'(t) \in \partial\ \varphi^*(t,\ X(t)) \quad \text{a.e.}$$

BIBLIOGRAPHY OF CHAPTER VII

1 - BERGE, C. : Espaces topologiques. Fonctions multivoques. Dunod 1958.

2 - BOURBAKI, N. : Intégration des mesures. Chapitre 5 2ème édition. Paris. Hermann Paris

3 - BOURBAKI, N. : Intégration sur les espaces topologiques séparés. Chapitre 9. Hermann Paris 1969.

4 - CASTAING, Ch. : Intégrales convexes duales. C.R. Acad. Sc. Paris 275, 1331-1334 (1972) et Exposé N°6, Séminaire d'Analyse Convexe Montpellier (1973).

5 - CASTAING, Ch. : Proximité et Mesurabilité. Un théorème de compacité faible. Colloque sur la théorie mathématique de contrôle optimal Bruxelles (1969).

6 - CASTAING, Ch. : Version aléatoire du problème de râfle par un convexe C.R. Acad. Sc. Paris 277, 1057-1059 (1973) et Exposé N°1, Séminaire d'Analyse convexe Montpellier (1974).

7 - GROTHENDIECK, A : Espaces vectoriels topologiques. Sao Paulo.

8 - HIMMELBERG, C.J., JACOB, M.Q., VAN VLECK, F.S. : Measurable multi-functions, selections and Filippov's implicit function lemma, J. Math. Analysis Applications 25-2, 276-284 (1969).

9 - IOFFE, A.D., TIKHOMIROV, V.N. : Duality of convex functions and extremum problems, Uspekhi Mat. Nauk 23,51-116 (1968).

10 - LINDENTRAUSS, J. : On non separable reflexive Banach space, Bull. Amer. Math. Soc. 72, 967-970 (1966).

11 - MICHAEL, E. : Continuous selections I, Ann. of Math. 63, 361-382 (1956).

12 - MOKOBODZKI, G. : Barycentres généralisés, Séminaire Brelot Choquet Deny, Faculté des Sciences Paris (1962).

13 - MOREAU, J.J. : Proximité et dualité dans un espace hilbertien. Bull. Soc. Math. France 93, 273-299 (1965).

14 - MOREAU, J.J. : Fonctionnelles convexes, Séminaire sur les équations aux dérivées partielles II, Collège de France (1966-1967).

15 - MOREAU, J.J. : Râfle par un convexe (Première partie), Exposé n°15, Séminaire d'Analyse convexe Montpellier (1971).

16 - MOREAU, J.J. : Râfle par un convexe variable (Deuxième partie), Exposé n°3, Séminaire d'Analyse convexe Montpellier (1972) et C.R. Acad. Sc. Paris 276, 791-794 (1973).

17 - MOREAU, J.J. : Rétraction d'une multi-application, Exposé n°13, Séminaire d'Analyse convexe Montpellier (1973) et C.R. Acad. Sc. Paris 276, 265-268 (1973).

18 - NEVEU, J. : Bases mathématiques du Calcul des Probabilités, Masson Editeur 1964.

19 - PERALBA, J.C. : Equations d'Evolution dans un espace de Hilbert associés à des opérateurs sous-différentiels, Thèse 3ème cycle, Montpellier 1972-1973.

20 - ROCKAFELLAR, R.T. :Integrals which are convex functionals. Pacific Journal of Math 24-3, 525-539 (1968).

21 - ROCKAFELLAR, R.T. : Integrals which are convex functionals II. Pacific Journal of Math 39-2, 439-469 (1971).

22 - ROCKAFELLAR, R.T. : Convex integral functionals and duality. Contribution to non linear functional analysis. Academic Press, 215-236 (1971).

23 - STRASSEN, V. : The existence of probability measures with given marginals. Ann. Math. Stat. 38, 423-439 (1965).

24 - TULCEA, C. : Two theorems concerning disintegration of measures. J. Math. Analysis Appl. 26, 376-380 (1969).

25 - TULCEA, A., TULCEA, C. : Topics in theory of liftings. Springer-Verlag. Berlin Heidelberg New York 1969.

26 - VALADIER, M. : Désintégration d'une mesure sur un produit C.R. Acad. Sc. Paris 276, 33-35 (1973) et Exposé n°10, Séminaire d'Analyse convexe Montpellier (1972).

27 - VALADIER, M. : Sur le théorème de Strassen. C.R. Acad. Sc. Paris, 278 1021-1024 , (1974) et Exposé n°4, Séminaire d'Analyse convexe Montpellier (1974).

28 - VALADIER, M. : Integrandes sur les localement convexes sous liniens. C.R. Acad. Sc. Paris 276, 693-695 (1973) et Exposé n°2, Séminaire d'Analyse convexe Montpellier (1973).

CHAPTER VIII

A NATURAL SUPPLEMENT OF L^1 IN THE DUAL OF L^∞. APPLICATIONS.

This chapter contains a detailed proof of the decomposition theorem of the dual space of L_E^∞, and applications of that theorem to conditional expectations of integrands and random sets.

We prove the decomposition theorem under rather general hypotheses. Many readers may be interested in the case where E is a reflexive separable Banach space. They will skip numbers 2 and 3 which study the special space $L_{E'}^1[E]$. Readers only interested by the case when μ is bounded or σ-finite will read prop.9 after definition 5.

The decomposition property is probably well known to specialists in Banach algebras when $E = \mathbb{R}$, and it is a consequence of Yosida-Hewitt [31]. We give a proof using stonian spaces (§3) which reduces it to the Lebesgue decomposition theorem for measures. We give also another proof (§4) following DubovitskiiMiliutin [8] which uses Riesz space properties. In § 5) the result is extended to the case for E Banach. The method is that of Pallu de La Barrière [20] who gave the best and shortest proof, using a kind of total variation. Many papers have been published on this question : Fhima [10], Ioffe-Levin [13], Levin [16] [17]. Rockafellar [23] and Valadier [27] have noticed a small incompleteness in the proof of [13]. Finally Yosida-Hewitt [31] and Huff [12] (see also his bibliography) study the decomposition of finitely additive measures.

As an application we give first a general theorem due to Rockafellar [24] (§7). Then we apply that result to conditional expectations. We follow partly Bismut [1], partly Valadier [28]. As integrals are particular cases of conditional expectations, we give in a corollary a result (corollary 39) on integration of multifunctions which extends in one direction the

Strassen's theorem V-14.

§1 - SINGULAR LINEAR FUNCTIONALS ON L^∞_E. STATEMENT OF THE MAIN THEOREM.

1 - Let $(T, \mathcal{C}, \mu)$ be a measured space : that is T is a set, $\mathcal{C}$ a tribe of subsets of T, and μ a positive measure. We suppose that $\mathcal{C}$ is complete and that $(T, \mathcal{C}, \mu)$ has the following direct sum property : there exists a family $(T_k)_{k \in K}$ with the properties

i) $T_k \in \mathcal{C}$

ii) $k \neq k' \Rightarrow T_k \cap T_{k'} = \emptyset$

iii) $\mu(T_k) < \infty$

iv) $A \subset T$ and $(\forall k)\ A \cap T_k \in \mathcal{C} \Rightarrow A \in \mathcal{C}$

v) $\forall A \in \mathcal{C},\ \mu(A) = \sum_{k \in K} \mu(A \cap T_k)$.

That property is verified when μ is bounded or σ-finite, and also when T is a Hausdorff topological space and μ the essential measure of a Radon measure on T (then the T_k may be chosen compact and the family locally countable ; that is a "concassage" : see Bourbaki [2]). That property is equivalent to saying that μ is strictly localizable (Dinculeanu [6], Ionescu Tulcea [15], Zaanen [32]). Therefore $L^\infty_{\mathbb{R}}(T, \mathcal{C}, \mu)$ is order complete, and every family $(A_i)_{i \in I}$ in $\mathcal{C}$ has an essential supremum A (that is $(\forall i)$, $A_i - A$ is negligible, and if $B \in \mathcal{C}$ is such that $(\forall i)$ $A_i - B$ is negligible, then so is $A-B$). Recall that if μ is σ-finite there exists a countable subfamily of $(A_i)_{i \in I}$ which has the same essential supremum.

2 - Let E be a Banach space. We denote by $\mathcal{L}^\infty_E(T, \mathcal{C}, \mu)$ the vector space of bounded functions whose restrictions to every T_k are strongly measurable (that is $\varphi_{|T_k}$ is the limit almost everywhere of a sequence of measurable functions assuming a finite number of values). Two functions φ, ψ are equivalent if $\varphi = \psi$ a.e. The quotient space, $L^\infty_E(T, \mathcal{C}, \mu)$, is a Banach space with the norm

$\|\tilde{\varphi}\|_\infty = \text{ess sup}\{\|\varphi(t)\| \mid t \in T\}$ (here $\tilde{\varphi}$ denotes the equivalence class of φ).

We denote by $\mathcal{L}^1_{E'}[E]$ the vector space of functions $f : T \to E'$ such that

- $\forall x \in E$, $< f(.),x >$ is measurable (we say that f is scalarly measurable)
- and there exists an integrable function h such that $\|f(t)\| \leq h(t)$.

A semi-norm on $\mathcal{L}^1_{E'}[E]$ is defined by

$$N_1(f) = \int^* \|f(t)\|\mu(dt) = \inf\{\int h\ \mu \mid h \in \mathcal{L}^1_{\mathbb{R}}\ ,\ h \geq \|f\|\}.$$

Two functions $f,g \in \mathcal{L}^1_{E'}[E]$ are equivalent if $\forall x \in E$, $< f,x > = < g,x >$ a.e. (we say that $f = g$ scalarly a.e.). The equivalence class of f is denoted by $\dot{f}$. The quotient space $L^1_{E'}[E]$ is normed by

$\|\dot{f}\|_1 = \inf\{N_1(g) \mid g \in \dot{f}\}$. Indeed : if $\|\dot{f}\|_1 = 0$, for every $\varepsilon > 0$ there exists $g \in \dot{f}$ such that $\int^* \|g\|\mu \leq \varepsilon$. For every x

$|<f,x>| = |<g,x>|$ a.e. and

$$\int |<f,x>|\mu = \int |<g,x>|\mu \leq \|x\| \int^* \|g\|\mu \leq \varepsilon\|x\|$$

Hence $<f,x> = 0$ a.e.

We shall obtain as corollary of the main theorem that $L^1_{E'}[E]$ is a Banach space.

3 - <u>Lemma VIII.3.</u> - <u>The space</u> $L^1_{E'}[E]$ <u>is isometric to a subspace of</u> $(L^\infty_E)'$.

<u>Moreover for every</u> $f \in \mathcal{L}^1_{E'}[E]$, <u>there exists</u> $g \in \dot{f}$ <u>such that</u> $\|g(.)\|$ <u>is measurable and</u> $\|\dot{f}\|_1 = \int \|g(t)\|\mu(dt)$.

<u>Proof.</u>

1) First we have to define a pairing between $L^1_{E'}[E]$ and L^∞_E.

Let $f \in \mathcal{L}^1_{E'}[E]$ and $\varphi \in \mathcal{L}^\infty_E$. Then $t \mapsto <f(t),\varphi(t)>$ is measurable (thanks to the definition of $\mathcal{L}^\infty_E$), and

$|<f(t), \varphi(t)>| \leq \|\varphi\|_\infty \|f(t)\|$ a.e., so that $t \mapsto < f(t), \varphi(t) >$ is integrable.

If $g \in \dot{f}$ and $\psi \in \tilde{\varphi}$ then $< f(t),\varphi(t)> = < f(t), \psi(t) >$ a.e. and $< f(t),\varphi(t) > = < g(t),\varphi(t) >$ a.e. (because on every T_k, φ is a limit almost everywhere of a sequence of measurable functions assuming a finite number of values).

So $\int < f(t),\varphi(t) > \mu\,(dt)$ depends only on the classes $\dot{f}$ and $\tilde{\varphi}$.

We put $< \dot{f},\tilde{\varphi} > = \int < f(t),\varphi(t) > \mu(dt)$.

2) As for every $g \in \dot{f}$

$$< \dot{f},\tilde{\varphi} > = \int < g(t),\varphi(t) > \mu(dt) \leq \|\tilde{\varphi}\|_\infty \; N_1(g)$$

one has $< \dot{f},\tilde{\varphi} > \leq \|\tilde{\varphi}\|_\infty \, \|\dot{f}\|_1$.

3) It remains to prove the inequality

$$\sup\{< \dot{f},\tilde{\varphi} > \mid \|\tilde{\varphi}\|_\infty \leq 1\} \geq \|\dot{f}\|_1 .$$

a) Let ρ be a lifting of $L^\infty_{\mathbb{R}}$. Let $f \in \mathcal{L}^1_{E'}[E]$.

Let $h \in \mathcal{L}^1_{\mathbb{R}}$ such that $h \geq \|f\|$ and $A_n = \{t \mid n \leq h(t) < n+1\}$.

We may modify h on a negligible set to obtain $A_n = \rho(A_n)$. Thus we have only $\|f\| \leq h$ a.e.

For every n, let $\varphi^n_x \in L^\infty_{\mathbb{R}}$ denote the set of measurable bounded functions equal a.e. to $\chi_{A_n} < f,x >$.

Then $x \mapsto \varphi^n_x$ is linear and $\|\varphi^n_x\|_\infty \leq \|x\| \, (n+1)$.

So $x \mapsto \rho(\varphi^n_x)\,(t)$ defines an element $g^n(t) \in E'$, and one has $g^n(t) = 0$ if $t \notin A_n$.

Put $g(t) = \begin{cases} g^n(t) & \text{if } t \in A_n \\ 0 & \text{if } t \in T - \bigcup A_n . \end{cases}$

Then $g = f$ scalarly almost everywhere.

Moreover on every A_n, as $\varphi^n_x \leq \|x\| \chi_{A_n} h$

$$< g^n(t), x > \leq \|x\| \; \rho\,(\chi_{A_n} h)\,(t).$$

We put $h'(t) = \begin{cases} \rho(\chi_{A_n} h)\,(t) & \text{if } t \in A_n \\ 0 & \text{if } t \in T - \bigcup A_n . \end{cases}$

As $h' = h$ a.e. h' is integrable, and $\|g(t)\| \leq h'(t)$ proves that g belongs to $\mathcal{L}^1_{E'}[E]$.

b) Consider now the set Ψ of functions $\varphi = \sum_{i=1}^{n} x_i \chi_{B_i}$ such that the B_i are pairwise disjoint, $B_i \in \mathcal{C}$, $\|x_i\| \leq 1$, and ϕ the subset of Ψ defined by the supplementary condition $\rho(B_i) = B_i$. We shall use the family $(<g(.),\varphi(.)>)_{\varphi \in \phi}$.

Remark that, for $\varphi \in \phi$,

$$\begin{aligned}\rho(\chi_{A_n} <g(.),\varphi(.)>) &= \rho(\Sigma\, \chi_{A_n} <g(.),\, x_i > \chi_{B_i}) \\ &= \Sigma\, \chi_{A_n} < g(.),\, x_i > \chi_{B_i} \\ &= \chi_{A_n} < g(.),\varphi(.) >.\end{aligned}$$

Let us show that the family $(\chi_{A_n} < g(.),\varphi(.) >)_{\varphi \in \phi}$ is directed. Indeed if $\varphi, \varphi' \in \phi$ there exists $\psi \in \Psi$, $\psi = \Sigma\, z_i \chi_{C_i}$ such that $< g,\psi > \geq \sup\, (< g,\varphi >, < g,\varphi' >)$. We may replace ψ by $\rho(\psi) = \Sigma\, z_i \chi_{\rho(C_i)}$. As $< g,\rho(\psi) > = < g,\psi >$ a.e. and as $\chi_{A_n} < g,\rho(\psi) >, \chi_{A_n} <g,\varphi >, \chi_{A_n} <g,\varphi'>$ are invariant by ρ we have still $< g,\rho(\psi) > \geq \sup\, (< g,\varphi >,< g,\varphi'>)$ on A_n. By a theorem of Ionescu Tulcea [15] (p. 40) $\sup_{\varphi \in \phi} \chi_{A_n} < g,\varphi >$ is measurable and $\int_{A_n} \sup_{\varphi \in \phi} < g,\varphi >\mu = \sup_{\varphi \in \phi} \int_{A_n} <g,\varphi>\mu$.

As ϕ contains the constant functions $x\, \chi_T$ (with $\|x\| \leq 1$) we have $\sup_{\varphi \in \phi} < g,\varphi > = \|g\|$. Thus $\|g(.)\|$ is measurable.

Then

$$\begin{aligned}\sup_{\varphi \in \phi} \int < g,\varphi > &= \sup_{\varphi}\, \Sigma \int_{A_n} < g,\varphi > \\ &= \Sigma \int_{A_n} \sup_{\varphi} < g,\varphi > \\ &= \int \|g\| \geq \|\dot{f}\|_1 .\end{aligned}$$

That proves the required inequality, and obviously $\int \|g\| d\mu = \|\dot{f}\|_1$.

Remark.

The space $\mathscr{L}^1_{E'}[E]$ is defined in Ionescu Tulcea [15]. Another definition is natural. Let $\bar{\mathscr{L}}^1_{E'}[E]$ denotes the space of scalarly measurable functions $f : T \to E'$ such that $\operatorname{ess\,sup}_{\|x\|\leq 1} < f(.),x >$ is integrable, with the semi-norm

$$\bar{N}_1(f) = \int [\operatorname{ess\,sup}_{\|x\| \leq 1} < f,x >]\mu .$$

We say that f and g are equivalent if $f = g$ scalarly almost everywhere. The semi-norm is constant on every equivalence class, and $\bar{N}_1(f) = 0 \Leftrightarrow f = 0$ scalarly a.e. So the quotient space $\bar{L}^1_{E'}[E]$ is normed by $|||\bar{f}|||_1 = \bar{N}_1(f)$ (here $\bar{f}$ denotes the equivalence class of f). Fortunately the spaces $L^1_{E'}[E]$ and $\bar{L}^1_{E'}[E]$ are isometric. Indeed one can prove first that if $f \in \bar{\mathscr{L}}^1_{E'}[E]$ there exists $g \in \mathscr{L}^1_{E'}[E] \cap \bar{f}$ (the function g obtained in 3) a) of the proof has that property). Finally the last part of the proof (3) b)) shows that if $f \in \mathscr{L}^1_{E'}[E]$ $\|\dot{f}\|_1 = |||\bar{f}|||_1$.

4 - Definition - Let $\ell \in (L^\infty_E)'$ and $A \in \mathscr{C}$. We shall say that ℓ is supported by A if $\varphi \in L^\infty_E$, $\varphi|_A = 0$ implies $\ell(\varphi) = 0$.

5 - Definition - Let $\ell \in (L^\infty_E)'$. We shall say that ℓ is singular if there exists a family $(A_i)_{i\in I}$ in $\mathscr{C}$ such that $(\forall i)$ ℓ is supported by A_i, and ess inf $A_i = \emptyset$ (we shall say that ℓ is (A_i)-singular).

We shall denote by L_s the set of singular linear continuous functionals on L^∞_E. We shall prove later (prop.7) that L_s is a vector space.

Main theorem VIII.5. - The Banach space $(L^\infty_E)'$ is isomorphic and isometric to the sum $L^1_{E'}[E] \oplus L_s$ endowed with the norm $\|\dot{f}+\ell\| = \|\dot{f}\|_1 + \|\ell\|$.

Corollary. - The normed space $L^1_{E'}[E]$ is a Banach space.

We give now two examples of singular functionals.

Example 1. Suppose that $L^1_{\mathbb{R}}(T,\mathcal{C},\mu)$ has infinite dimension. Then μ has an infinity of atoms or has a non null non atomic part. In both cases there exists a family (A_n) in $\mathcal{C}$ such that (A_n) is decreasing, $\bigcap A_n = \emptyset$ and $0 < \mu(A_n) < \infty$. Put

$$\ell_n(\tilde{\varphi}) = \frac{1}{\mu(A_n)} \int_{A_n} \varphi(t)\ \mu(dt) \quad \text{for every } \tilde{\varphi} \in L^\infty_{\mathbb{R}}.$$

Let $\mathcal{U}$ be an ultrafilter on $\mathbb{N}$ finer than the Fréchet filter. Then $\ell(\tilde{\varphi}) = \lim \ell_n(\tilde{\varphi})$ defines a linear functional on $L^\infty_{\mathbb{R}}$. One has $|\ell(\tilde{\varphi})| \leq \|\tilde{\varphi}\|_\infty$ because $\|\ell_n\| = 1$, and $\ell(\chi_T) = 1$, so that ℓ is continuous and $\ell \neq 0$.

And for every n, ℓ is supported by A_n. Hence ℓ is singular.

Example 2. Let I be an infinite set, and μ the measure on $(I,\mathcal{P}(I))$ defined by

$$\mu(A) = \begin{cases} \text{Card}\ (A) & \text{if } A \text{ is finite} \\ +\infty & \text{otherwise} \end{cases}$$

Then $L^\infty_{\mathbb{R}}(I,\mathcal{P}(I),\mu) = \ell^\infty_{\mathbb{R}}(I)$. Let $\mathcal{U}$ be an ultrafilter on I finer than the filter of complementaries of finite subsets of I.

And put for every φ, $\ell(\varphi) = \lim\limits_{\mathcal{U}} \varphi(i)$. Then ℓ is linear continuous and $\ell(\chi_I) = 1$. Let $\mathcal{P}_f(I)$ denote the set of finite subsets of I. For every $A \in \mathcal{P}_f(I)$ ℓ is supported by $I-A$. As ess $\inf\{I-A \mid A \in \mathcal{P}_f(I)\} = \emptyset$, ℓ is singular.

6 - Lemma VIII.6. - If $\ell \in (L^\infty_E)'$ is supported by A and by B, then ℓ is supported by $A \cap B$.

Proof. Let $\varphi \in L^\infty_E$ and suppose that $\varphi_{|A\cap B} = 0$. Then $\ell(\varphi) = \ell(\chi_A\varphi)+\ell(\chi_{T-A}\varphi)$ and $\chi_A\varphi$ is zero on B and $\chi_{T-A}\varphi$ is zero on A. So $\ell(\varphi) = 0$.

Remark. In definition 5 the family $(A_i)_{i\in I}$ may be replaced by a directed family $(B_u)_{u\in U}$. Indeed let $U = \mathcal{P}_f(I)$ and for every $u \in U$ put

$B_u = \bigcap_{i \in u} A_i$.

7 - Proposition VIII.7. - The set of singular functionals L_s is a vector space and $L^1_{E'}[E] \cap L_s = \{0\}$.

Proof.

1) It is obvious that 0 is singular (it is supported by ϕ).

If ℓ is (A_i)-singular and if $\lambda \in \mathbb{R}$, then $\lambda\ell$ is still (A_i)-singular. If in addition m is (B_j)-singular, let us prove that $\ell+m$ is singular. For every $(i,j) \in I \times J$, $\ell+m$ is supported by $A_i \cup B_j$.

And
$$\operatorname*{ess\,inf}_{(i,j)} (A_i \cup B_j) = \operatorname*{ess\,inf}_i [\operatorname*{ess\,inf}_j (A_i \cup B_j)]$$
$$= \operatorname*{ess\,inf} A_i = \phi.$$

2) Let $\dot{f} \in L^1_{E'}[E] \cap L_s$. There exists a family $(A_i)_{i \in I}$ such that $\dot{f}$ is (A_i) singular. Then if $\tilde{\varphi} \in L^\infty_E$ and $i \in I, \langle \dot{f}, \chi_{T-A_i} \tilde{\varphi} \rangle = 0$. Therefore $f(t) = 0$ scalarly a.e. on $T-A_i$. So if $x \in E$, $\{t \mid \langle f(t),x\rangle = 0\}$ contains $T - A_i$ except a negligible set. As ess sup $(T-A_i) = T$, it follows that $f=0$.

Remark. The equality $L^1_{E'}[E] \cap L_s = \{0\}$ is an easy corollary of the following proposition. Indeed, if $\ell \in L^1_{E'}[E] \cap L_s$, then

$$2\|\ell\| = \|\ell\| + \|-\ell\| = \|\ell+(-\ell)\| = 0.$$

8 - Proposition VIII.8. - If $\dot{f} \in L^1_{E'}[E]$ and $\ell \in L_s$, then $\|\dot{f}+\ell\| = \|\dot{f}\|_1 + \|\ell\|$.

Proof. We have to prove $\|\dot{f}+\ell\| \geq \|\dot{f}\| + \|\ell\|$. Let us suppose that ℓ is (A_i)- singular, and suppose that the family (A_i) is directed (see remark in n°6). We may suppose by lemma 3 that $\|f(.)\|$ is measurable and that $\|\dot{f}\|_1 = \int \|f(t)\|\mu\,(dt)$. As (A_i) is directed for every $\varepsilon > 0$ there exists i such that $\int_{T-A_i} \|f(t)\|\mu\,(dt) \geq \|\dot{f}\|_1 - \varepsilon$ (indeed the set $B=\{t \mid \|f(t)\|>0\}$ is σ-finite, and there exists a sequence (i_n) such that the sequence

A_{i_n} decreases and that $\cap A_{i_n} \cap B$ is negligible). Let $\varphi_1 \in \mathcal{L}^\infty_E$ such that $\|\tilde{\varphi}_1\|_\infty \le 1$ and $\ell(\varphi_1) \ge \|\ell\| - \varepsilon$. We may suppose $\varphi_1(t) = 0$ on $T - A_i$. There exists $\varphi_2 \in \mathcal{L}^\infty_E$ such that $\|\tilde{\varphi}_2\|_\infty \le 1$, $\varphi_2(t) = 0$ on A_i and $< f,\varphi_2 > \ge \int_{T-A_i} \|f(t)\|\mu(dt) - \varepsilon$: indeed if $\psi \in \mathcal{L}^\infty_E$ is such that $\|\tilde{\psi}\|_\infty \le 1$ and $< f,\psi > \ge N_1(\dot{f}) - \varepsilon = \int_T \|f(t)\|\mu(dt) - \varepsilon$,

put $\varphi_2 = \chi_{T-A_i}\ \psi$. Then

$$< f,\chi_{A_i}\psi > \le \int_{A_i} \|f(t)\|\mu(dt),$$

hence $< f,\varphi_2 > = < f,\psi - \chi_{A_i}\psi > \ge \int_{T-A_i} \|f(t)\|\mu(dt) - \varepsilon$.

Hence

$$\begin{aligned} < f+\ell,\varphi_1+\varphi_2 > &= < f,\varphi_1 > + < f,\varphi_2 > + < \ell,\varphi_1 > \\ &\ge < f,\varphi_1 > + \|\dot{f}\|_1 - 2\varepsilon + \|\ell\| - \varepsilon \\ &\ge \|\dot{f}\|_1 + \|\ell\| - 4\varepsilon. \end{aligned}$$

As $\|\tilde{\varphi}_1 + \tilde{\varphi}_2\|_\infty \le 1$ that proves $\|\dot{f}+\ell\| \ge \|\dot{f}\|_1 + \|\ell\|$.

<u>Remark</u>. To prove the main theorem it remains (thanks to prop. 7 and 8) to prove the equality $(L^\infty_E)' = L^1_{E'}[E] + L_s$.

9 - <u>Proposition VIII.9</u>. - <u>Let</u> $\ell \in (L^\infty_E)'$

1) <u>If</u> μ <u>is</u> σ-<u>finite</u> "ℓ <u>is singular</u>" <u>is equivalent to "there exists a decreasing sequence</u> $(B_p)_{p \in \mathbb{N}}$ <u>in</u> $\mathcal{T}$ <u>such that</u> ℓ <u>is</u> (B_p)-<u>singular</u> (<u>that is</u> $\cap B_p$ <u>is negligible and</u> ℓ <u>is supported by every</u> B_p)".

2) <u>If</u> μ <u>is bounded</u> "ℓ <u>is singular</u>" <u>is equivalent to</u>" $\forall\ \varepsilon > 0$ <u>there exists</u> $A \in \mathcal{T}$, <u>such that</u> $\mu(A) \le \varepsilon$ <u>and</u> ℓ <u>is supported by</u> A".

<u>Proof</u>.

1) Suppose that ℓ is (A_i)-singular. We may suppose the family (A_i) directed (remark 6). Therefore there exists a sequence (i_n) such that $B_n = A_{i_n}$ is decreasing and $\cap B_n$ is negligible.

2) The implication $\Rightarrow$ follows from 1). Conversely if for every $\varepsilon > 0$ there exists A_ε such that $\mu(A_\varepsilon) \leq \varepsilon$ and ℓ is supported by A_ε, then ℓ is $(A_\varepsilon)_{\varepsilon>0}$-singular

10 - Proposition VIII.10. - Let $\ell \in (L_E^\infty)'$. For $A \in \mathcal{C}$ we define $\ell^A \in (L_E^\infty)'$ by $\ell^A(\tilde{\varphi}) = \ell(\tilde{\chi}_A \tilde{\varphi})$. Then if $A_1 \cap A_2 = \emptyset$ one has $\|\ell^{A_1 \cup A_2}\| = \|\ell^{A_1}\| + \|\ell^{A_2}\|$

Proof. We have to prove $\|\ell^{A_1 \cup A_2}\| \geq \|\ell^{A_1}\| + \|\ell^{A_2}\|$.

Let $\varphi_i \in \mathcal{L}_E^\infty (i = 1,2)$ be such that $\|\tilde{\varphi}_i\|_\infty \leq 1$ and $\ell^{A_i}(\tilde{\varphi}_i) \geq \|\ell^{A_i}\| - \varepsilon$

Put

$$\varphi(t) = \begin{cases} \varphi_1(t) & \text{if } t \in A_1 \\ \varphi_2(t) & \text{if } t \in T - A_1 \end{cases}$$

Then

$$\begin{aligned} \ell^{A_1 \cup A_2}(\varphi) &= \ell(\chi_{A_1 \cup A_2} \varphi) \\ &= \ell(\chi_{A_1} \varphi + \chi_{A_2} \varphi) \\ &= \ell^{A_1}(\varphi_1) + \ell^{A_2}(\varphi_2) \\ &\geq \|\ell^{A_1}\| + \|\ell^{A_2}\| - 2\varepsilon . \end{aligned}$$

As $\|\tilde{\varphi}\|_\infty \leq 1$, that proves the required inequality.

11 - The property in the following lemma has been taken as definition of singularness by Dubovitskii-Miliutin [8].

Lemma VIII.11. - Let $\ell \in (L_E^\infty)'$. If for every $\varepsilon > 0$ there exists $A \in \mathcal{C}$ such that $\mu(A) \leq \varepsilon$ and $\|\ell^A\| \geq \|\ell\| - \varepsilon$, then ℓ is singular.

Proof. Let A_n be such that $\mu(A_n) \leq 2^{-n}$ and $\|\ell^{A_n}\| \geq \|\ell\| - 2^{-n}$.

Then, thanks to prop. 10, $\|\ell^{T-A_n}\| \leq 2^{-n}$. Put $B_n = \bigcup_{k \geq n+1} A_k$. Then $\mu(B_n) \leq 2^{-n}$. Let us show that μ is supported by B_n. If $\varphi \in \mathcal{L}_E^\infty$, $\varphi(t) = 0$ on B_n, then for $k \geq n+1$, $\varphi = \chi_{T-A_k} \varphi$. Hence

$$|\ell(\varphi)| = |\ell^{T-A_k}(\varphi)| \le \|\ell^{T-A_k}\| \, \|\tilde{\varphi}\|_\infty \le 2^{-k}\|\varphi\|.$$

Thus $\ell(\varphi) = 0$, and ℓ is (B_n)-singular.

12 - We do not need the following corollary, but it is easy to give it now. **The result is** useful in the proofs of Ioffe-Levin [13] and Fhima [10].

Corollary VIII 12. - We suppose μ bounded. Denote by $L^\infty_{\mathbb{R}} \otimes E$ the subspace of L^∞_E generated by the set of functions $\{u\, x \mid u \in L^\infty_{\mathbb{R}},\ x \in E\}$. Let $\ell \in (L^\infty_E)'$. If for every $\varphi \in L^\infty_{\mathbb{R}} \otimes E$, $\ell(\varphi) = 0$, then ℓ is singular.

Proof. Let H be the subspace of L^∞_E of the equivalence classes of functions assuming a countable set of values. As μ is bounded, H is dense in L^∞_E. Therefore there exists $\tilde{\varphi} \in H$ such that $\|\tilde{\varphi}\|_\infty \le 1$ and $\ell(\varphi) \ge \|\ell\| - \varepsilon$. We may suppose that there exists a sequence (T_p) in $\mathcal{T}$ and a sequence (x_p) in E such that $\varphi(t) = x_p$ if $t \in T_p$, $\cup T_p = T$ and the T_p are pairwise disjoint. Let n be large enough in order that

$\mu(\bigcup_{p\ge n} T_p) \le \varepsilon$. As $\sum_{p=0}^{n-1} \chi_{T_p} x_p$ belongs to $L^\infty_{\mathbb{R}} \otimes E$,

one has $\ell(\varphi) = \ell(\chi_{\bigcup_{p\ge n} T_p}\, \varphi) \ge \|\ell\| - \varepsilon$.

Thus the hypothesis of lemma 11 is verified.

§2 - REPRESENTATION OF $L^\infty_{\mathbb{R}}$. STONIAN SPACES.

We suppose in this section that μ is bounded.

13 - We refer to Bourbaki [3]. The space $L^\infty_{\mathbb{C}}(T, \mathcal{T}, \mu)$ is a Banach algebra with the involution $\varphi \mapsto \bar{\varphi}$. Its spectrum $\tilde{T}$ is compact, and the Gelfand transform $\mathcal{G} : L^\infty_{\mathbb{C}} \to \mathcal{C}_{\mathbb{C}}(\tilde{T})$ is an isomorphism. Moreover $\mathcal{G}(L^\infty_{\mathbb{R}}) = \mathcal{C}_{\mathbb{R}}(\tilde{T})$ because a function is real iff it is equal to its adjoint. Moreover if $\varphi \in L^\infty_{\mathbb{R}}$, $\varphi \ge 0$ is equivalent to $\mathcal{G}(\varphi) \ge 0$. Indeed if $\varphi \in L^\infty_{\mathbb{R}}$ (resp. $\tilde{\varphi} \in \mathcal{C}_{\mathbb{R}}(\tilde{T})$), $\varphi \ge 0$ (resp. $\tilde{\varphi} \ge 0$) is equivalent to $\forall \varepsilon > 0$, $\varphi + \varepsilon\, \chi_T$ (resp. $\tilde{\varphi} + \varepsilon \chi_{\tilde{T}}$) has an inverse. (The isomorphism of $L^\infty_{\mathbb{R}}(T)$ with the

space of continuous functions on a compact set is well known: see Choquet [4] Dunford-Schwartz [9] p. 395). So $\widetilde{T}$ is stonian, according to the following :

Definition. *A topological space* S, *is said to be stonian if it is compact, and if the ordered space* $\mathcal{C}_{\mathbb{R}}(S)$ *is order complete.*

We can remark that $\widetilde{T}$ has many open-closed subsets. Indeed if $A \in \mathcal{C}$, $x_A^2 = x_A$ implies that $\mathcal{G}(x_A)$ is a characteristic function. But if a characteristic function is continuous, it is the characteristic function of an open-closed set.

Example. Let I be a set. Then the spectrum of $\ell^{\infty}_{\mathbb{C}}(I)$, $\widetilde{I}$, is the Stone-Cech compactification of I (here the corresponding measure is not bounded but we have not used this hypothesis so far). In this particular case $\widetilde{I}$ is also the set of ultrafilters on I with a certain topology. A continuous linear form on $\ell^{\infty}_{\mathbb{R}}(I)$ is a real measure on $\widetilde{I}$. It is the sum of a measure supported by I (which is open in I) and of a measure supported by $\widetilde{I} - I$.

14 - Proposition VIII.14. - *Let* S *be stonian and* $(f_i)_{i \in I}$ *an upper bounded family in* $\mathcal{C}_{\mathbb{R}}(S)$. *Let* f *be the supremum of* (f_i) *in* $\mathcal{C}_{\mathbb{R}}(S)$ *and* g *the function defined by* $g(s) = \sup f_i(s)$. *Then* $\{s | f(s) \neq g(s)\}$ *is meagre.*

Proof. One has $f \geq g$. Put $S_n = \{s \in S | f(s) \geq g(s) + \frac{1}{n}\}$.
Then S_n is closed because f is continuous and g l.s.c..
Suppose that $\overset{\circ}{S}_n \neq \emptyset$. As S is compact there exists $h : S \to [0,1]$ continuous which is zero on $S - S_n$ and non identically zero on S_n.
But $f - \frac{1}{n} h$ would be $\geq g$, so that f would not be the supremum of the family (f_i). Thus $\overset{\circ}{S}_n = \emptyset$, and $\{s | f(s) \neq g(s)\} = \bigcup_{n \geq 1} S_n$ is meagre.

15 - Proposition VIII.15. - Let S be stonian, g be a l.s.c. bounded function on S. Then there exists a unique continuous function $\bar{g}$ such that $\{s|g(s) \neq \bar{g}(s)\}$ is meagre. Moreover $\bar{g}$ is the smallest u.s.c. function greater than g.

Proof. As S is compact there exists a family (f_i) in $\mathcal{C}_{\mathbb{R}}(S)$ such that $g(s) = \sup f_i(s)$. Let $\bar{g}$ denote the supremum of (f_i) in $\mathcal{C}_{\mathbb{R}}(S)$. By prop. 14 the set $\{s|g(s) \neq \bar{g}(s)\}$ is meagre. Thus $\bar{g}$ has the required property. Moreover as S is a Baire space, the complement of a meagre set is dense. Hence there is at most one continuous function h such that $\{s|g(s) \neq h(s)\}$ is meagre.

Finally let $D = \{s \in S|g(s) = \bar{g}(s)\}$. As D is dense, for every $s_o \in S$ one has

$$\bar{g}(s_o) = \lim_{\substack{s \to s_o \\ s \in D}} g(s), \text{ and as } \bar{g} \geq g \text{ one has}$$

$$\bar{g}(s_o) = \overline{\lim_{\substack{s \to s_o \\ s \in S}}} g(s).$$ That proves that $\bar{g}$ is the smallest u.s.c. function greater that g.

16 - Corollary VIII.16. - Let U be an open set in S. Then $\bar{U}$ is still open. Every point has a basis of open-closed neighbourhoods.

Proof. First χ_U is a l.s.c. function and the smallest u.s.c. function greater than χ_U is $\chi_{\bar{U}}$.

If V is a neighbourhood of $s_o \in S$, let W be a closed neighbourhood contained in V and U an open neighbourhood contained in W. Then $\bar{U}$ is an open-closed neighbourhood contained in V.

17 - Definition - Let S be stonian and ν be a positive Radon measure on S. Then ν is said to be normal if for every directed family $(f_i)_{i \in I}$ in $\mathcal{C}_{\mathbb{R}}(S)$ which is supposed upper bounded and whose supremum in $\mathcal{C}_{\mathbb{R}}(S)$ is

denoted by f, one has

$$\int f\nu = \sup_{i \in I} \int f_i\ \mu.$$

Example : We define a Radon measure $\tilde{\mu}$ on the spectrum $\tilde{T}$ of $L^\infty_{\mathbb{C}}(T,\mathcal{T},\mu)$ by $\tilde{\mu}(\tilde{\varphi}) = \int_T \mathcal{G}^{-1}(\tilde{\varphi})d\mu$ (recall that we have supposed μ bounded). Then $\tilde{\mu}$ is normal. Indeed for every upper bounded directed family $(\varphi_i)_{i\in I}$ in $L^\infty_{\mathbb{R}}(T)$ one has $\int (\text{ess sup } \varphi_i)\ d\mu = \sup \int \varphi_i\ d\mu$ (that follows from the properties of ess sup : there exists a sequence i_n such that (φ_{i_n}) is increasing and ess sup $\varphi_i = \sup \varphi_{i_n}$. Hence

$$\int (\text{ess sup } \varphi_i) = \int (\sup \varphi_{i_n}) = \sup \int \varphi_{i_n} \leq \sup \int \varphi_i \leq \int (\text{ess sup } \varphi_i)).$$

18 - Proposition VIII.18. - Let S be stonian and ν a positive Radon measure on S. Then the following properties are equivalent :

a) ν is normal

b) every meagre subset of S is ν-negligible.

Remark. We need only the implication a $\Rightarrow$ b, but we prove also b $\Rightarrow$ a for completeness.

Proof.

a) Suppose that ν is normal and that $\overset{\circ}{\bar{A}} = \emptyset$, and let us prove that $\nu(A) = 0$. The set H of functions f such that $f \in \mathcal{C}_{\mathbb{R}}(S)$ $f \geq \chi_{\bar{A}}$ is lower bounded and directed. As $\inf_{f\in H} f(s) = \chi_{\bar{A}}(s)$, and as $S-\bar{A}$ is dense in S, the infimum of H in $\mathcal{C}_{\mathbb{R}}(S)$ is 0. Thus $\nu(\bar{A}) = \inf_{f \in H} \int f\ d\nu$ (because ν is Radon) $= 0$ (because ν is normal).

b) Conversely if (f_i) is an upper bounded directed family in $\mathcal{C}_{\mathbb{R}}(S)$ and if f is its supremum in $\mathcal{C}_{\mathbb{R}}(S)$, let g denote the function $g(s) = \sup f_i(s)$. Then $\sup \int f_i\ d\nu = \int g\ d\nu$ (because ν is Radon)

$= \int f\ d\nu$ (thanks to the hypothesis and prop. 14).

19 - <u>Proposition VIII.19. - The Gelfand transform can be extended in a linear isometric bijective map</u> $\bar{\mathcal{G}} : L^1_{\mathbb{R}}(T,\mathcal{C},\mu) \to L^1_{\mathbb{R}}(\tilde{T},\tilde{\mu})$.

<u>If</u> $\varphi \in L^\infty_{\mathbb{R}}(T)$, $f \in L^1_{\mathbb{R}}(T)$ <u>one has</u>

$$\int_T f\varphi \, d\mu = \int_{\tilde{T}} \bar{\mathcal{G}}(f)\, \mathcal{G}(\varphi)\, d\tilde{\mu}.$$

<u>Proof</u>. The spaces $L^\infty_{\mathbb{R}}(T)$ and $\mathcal{C}_{\mathbb{R}}(\tilde{T})$ are dense respectively in $L^1_{\mathbb{R}}(T)$ and $L^1_{\mathbb{R}}(\tilde{T})$. And for $\varphi \in L^\infty_{\mathbb{R}}(T)$

$$\begin{aligned} \|\mathcal{G}(\varphi)\|_1 &= \int |\mathcal{G}(\varphi)| \, d\tilde{\mu} \\ &= \int \mathcal{G}(|\varphi|)\, d\tilde{\mu} \\ &= \int |\varphi| \, d\mu = \|\varphi\|_1 . \end{aligned}$$

Hence the first part of the statement is proved.

Finally if $\varphi \in L^\infty(T)$ the map $f \mapsto \int_{\tilde{T}} \bar{\mathcal{G}}(f)\mathcal{G}(\varphi)\, d\tilde{\mu} - \int_T f\varphi \, d\mu$ is continuous on $L^1(T)$ and is zero on $L^\infty(T)$ (because $\mathcal{G}(\varphi\psi) = \mathcal{G}(\varphi)\mathcal{G}(\psi)$ if φ, $\psi \in L^\infty(T)$), hence it is zero everywhere.

<u>Remark</u>. It follows from the proposition that every continuous linear form on $L^1(\tilde{T},\tilde{\mu})$ is given by a function $\tilde{\varphi} \in \mathcal{C}(\tilde{T})$. As the support of $\tilde{\mu}$ is $\tilde{T}$ we may write $\mathcal{C}(\tilde{T}) = L^\infty(\tilde{T},\tilde{\mu})$. More generally if S is stonian and if ν is a normal positive Radon measure on S, if $g \in \mathcal{L}^\infty_{\mathbb{R}}(S,\nu)$ there exists $f \in \mathcal{C}_{\mathbb{R}}(S)$ such that $f = g$ a.e. see Dixmier [7].

§3 - <u>FIRST PROOF OF THE MAIN THEOREM, WHEN $E = \mathbb{R}$ AND μ BOUNDED</u>

20 - For readers who **skipped** §2, recall that $\tilde{T}$ is a compact set, $\mathcal{G} : L^\infty_{\mathbb{C}}(T) \to \mathcal{C}_{\mathbb{C}}(\tilde{T})$ is an isomorphism, and that the measure $\tilde{\mu}$ on $\tilde{T}$ is defined by

$$\tilde{\mu}(\tilde{\varphi}) = \int_T \mathcal{G}^{-1}(\tilde{\varphi})\, d\mu \quad \text{for every } \tilde{\varphi} \in \mathcal{C}_{\mathbb{R}}(\tilde{T}).$$

Let $\ell \in (L^\infty_{\mathbb{R}}(T))'$. Let us define the Radon measure $\tilde{\ell}$ on $\tilde{T}$ by $\tilde{\ell}(\tilde{\varphi}) = \ell(\mathcal{G}^{-1}(\tilde{\varphi}))$. Let $\tilde{\ell} = \tilde{f}\,\tilde{\mu} + \tilde{\ell}_s$ be the Lebesgue decomposition of $\tilde{\ell}$: that is $\tilde{f} \in L^1(\tilde{T},\tilde{\mu})$ and $\tilde{\ell}_s$ is supported by a $\tilde{\mu}$-negligible set N.

Let $f = \bar{\mathcal{G}}^{-1}(\tilde{f})$ (see prop. 19). It remains to show that $\ell - f$ is singular. Let $\varepsilon > 0$. Let U be an open subset of $\widetilde{T}$ such that $N \subset U$ and $\widetilde{\mu}(U) \leq \varepsilon$. As $\bar{U} - U$ has empty interior, it is negligible (prop. 18). Hence $\bar{U}$ is an open-closed set (corollary 16) such that $N \subset \bar{U}$ and $\widetilde{\mu}(\bar{U}) \leq \varepsilon$. Let χ_A be an element in $\mathcal{G}^{-1}(\chi_{\bar{U}})$. One has $\mu(A) \leq \varepsilon$. If $\varphi \in L^\infty_{\mathbb{R}}(T)$ is zero on A, if $\widetilde{\varphi}$ denotes $\mathcal{G}(\varphi)$, then

$$
\begin{aligned}
< \ell-f,\varphi > &= < \ell,\varphi > - < f,\varphi > \\
&= < \widetilde{\ell},\widetilde{\varphi} > - < \widetilde{f},\widetilde{\varphi} > \quad \text{(prop. 19)} \\
&= < \widetilde{\ell}_s,\widetilde{\varphi} >.
\end{aligned}
$$

But $\widetilde{\varphi}\chi_{\bar{U}} = \mathcal{G}(\varphi\,\chi_A) = \mathcal{G}(0) = 0$. Hence $\widetilde{\varphi}$ is zero on N, and $< \ell-f,\varphi > = < \widetilde{\ell}_s,\widetilde{\varphi} > = 0$. Proposition 9 shows that $\ell-f$ is singular. By remark 8 the main theorem is proved.

§4 - SECOND PROOF OF THE MAIN THEOREM WHEN $E = \mathbb{R}$ AND μ BOUNDED

21 - For ordered vector spaces we refer to Bourbaki [2], Peressini [21], Vulikh [30].

Lemma VIII.21. - The space $(L^\infty_{\mathbb{R}})'$ is the space of order bounded linear functionals on $L^\infty_{\mathbb{R}}$.

Proof. First if $\ell \in (L^\infty_{\mathbb{R}})'$ let us prove that ℓ is order bounded. Indeed if $\psi \in L^\infty_{\mathbb{R}}$ and $\varphi \in L^\infty_{\mathbb{R}}$ with $|\varphi| \leq \psi$, one has

$$\|\varphi\|_\infty \leq \|\psi\|_\infty \text{ and } |\ell(\varphi)| \leq \|\ell\|\,\|\psi\|_\infty .$$

Conversely suppose that ℓ is order bounded. Then ℓ is bounded on the set $\{\varphi \in L^\infty_{\mathbb{R}} \mid -\chi_T \leq \varphi \leq \chi_T\}$. But that is the unit ball of $L^\infty_{\mathbb{R}}$. So ℓ is continuous.

Consequence. As $L^\infty_{\mathbb{R}}$ is a Riesz space (it is even order complete) $(L^\infty_{\mathbb{R}})'$ is an order complete Riesz space.

22 - Lemma VIII.22. - $L^1_{\mathbb{R}}$ is a band in $(L^\infty_{\mathbb{R}})'$.

Proof. Let $H \subset L^1_{\mathbb{R}}$, $H \neq \emptyset$, and $\ell \in (L^\infty_{\mathbb{R}})'$ which majorizes H.

We have to prove that the supremum of H in $(L^{\infty}_{\mathbb{R}})'$ belongs to $L^{1}_{\mathbb{R}}$.

We may suppose that H is directed, and that all the elements of H are positive (indeed let $f_0 \in H$, the set $H_0 = \{f \in H | f \geq f_0\}$ has the same supremum as H, and $\sup H_0 = \sup(H_0 - f_0) + f_0$).

Consider for every $A \in \mathcal{C}$

$$\tau(A) = \sup\{\int_A f\, d\mu \mid f \in H\}.$$

We have $0 \leq \tau(A) \leq \ell(\chi_A)$.

If $A \cap B = \phi$, clearly $\tau(A \cup B) \leq \tau(A) + \tau(B)$, but as H is directed, $\tau(A \cup B) = \tau(A) + \tau(B)$ (because if $\int_A f d\mu \geq \tau(A) - \epsilon$, $\int_B g\, d\mu \geq \tau(B) - \epsilon$, taking $h \geq \sup(f,g)$ one obtains

$$\tau(A \cup B) \geq \int_{A \cup B} h d\mu \geq \tau(A) + \tau(B) - 2\epsilon).$$

If (A_n) is an increasing sequence in $\mathcal{C}$, one has

$$\begin{aligned}\tau(\cup A_n) &= \sup\{\int_{\cup A_n} f\, d\mu \mid f \in H\}\\ &= \sup_f \sup_n \int_{A_n} f\, d\mu\\ &= \sup_n \sup_f \int_{A_n} f\, d\mu\\ &= \sup_n \tau(A_n).\end{aligned}$$

That proves that τ is a positive bounded measure. Furthermore $\mu(A) = 0$ implies $\tau(A) = 0$. Thus $\tau = h\mu$ with $h \in L^{1}_{\mathbb{R}}$. It is clear that h majorizes H, because for every $f \in H$ and every $A \in \mathcal{C}$, $\int_A h\, d\mu \geq \int_A f\, d\mu$.

Suppose that $m \in (L^{\infty}_{\mathbb{R}})'$ majorizes H. Then one has for every $\varphi \in L^{\infty}_{\mathbb{R}}$, $\varphi \geq 0$ and every $f \in H$

$$m(\varphi) \geq \int f\varphi\, d\mu.$$ Particularly

$$m(\chi_A) \geq \sup_{f \in H} \int_A f\, d\mu = \tau(A) = \int h\, \chi_A\, d\mu.$$

The inequality $m(\varphi) \geq \int h\varphi d\mu$ is valid for every $\varphi \geq 0$ assuming a finite number of values. By continuity the inequality holds for every $\varphi \geq 0$. Hence $h \leq m$, and h is the supremum of H.

23 - By lemma 22 and the Riesz theorem, $(L^\infty_{\mathbb{R}})'$ is the direct sum of $L^1_{\mathbb{R}}$ and of its complementary band. So if $\ell \in (L^\infty_{\mathbb{R}})'$, $\ell \geq 0$, one has $\ell = f + \ell_s$ with $f \in L^1_{\mathbb{R}}$, $f \geq 0$, $\ell_s \geq 0$ and $\inf(|g|, \ell_s) = 0$ for every $g \in L^1_{\mathbb{R}}$.

Lemma VIII.23 - Put $\ell_1 = \ell_s - \chi_T$, then

a) $\ell_s = \ell_1^+$ and $\chi_T = \ell_1^-$

b) $\|\ell_s - \chi_T\| = \|\ell_s\| + \|\chi_T\|$ (the norms are taken in $(L^\infty_{\mathbb{R}})'$).

Proof.

a) One has $\ell_s - \chi_T = \ell_1^+ - \ell_1^-$, hence $\ell_s - \ell_1^+ = \chi_T - \ell_1^-$. Let p denote $\ell_s - \ell_1^+$.

As $\ell_1^+ = \sup(\ell_1, 0) \leq \ell_s$, $p \geq 0$. So we have $p = \ell_s - \ell_1^+ = \chi_T - \ell_1^- \geq 0$.

Moreover $p \leq \ell_s$ and $p \leq \chi_T$.

As $\inf(\ell_s, \chi_T) = 0$, $p = 0$. Thus a) is proved.

b) We have to prove $\|\ell_1^+\| + \|\ell_1^-\| \leq \|\ell_1\|$. We have

$$\|\ell_1^+\| = \ell_1^+(\chi_T) = \sup\{\ell_1(\varphi_1) \mid \varphi_1 \in L^\infty_{\mathbb{R}},\ 0 \leq \varphi_1 \leq 1\}$$

$$\|\ell_1^-\| = \ell_1^-(\chi_T) = \sup\{-\ell_1(\varphi_2) \mid \varphi_2 \in L^\infty_{\mathbb{R}},\ 0 \leq \varphi_2 \leq 1\}.$$

So

$$\begin{aligned}\|\ell_1^+\| + \|\ell_1^-\| &= \sup_{0 \leq \varphi_1 \leq 1} \ell_1(\varphi_1) + \sup_{0 \leq \varphi_2 \leq 1} (-\ell_1(\varphi_2)) \\ &= \sup_{0 \leq \varphi_1 \leq 1} \ell_1(\varphi_1) + \sup_{-1 \leq \varphi_2' \leq 0} \ell_1(\varphi_2') \\ &\leq \sup_{-1 \leq \psi \leq 1} \ell_1(\psi) = \|\ell_1\|.\end{aligned}$$

24 - Proof of the main theorem

By remark 8 it remains to prove that if $\ell \in (L^\infty_{\mathbb{R}})'$, then $\ell = f + \ell_s$ with $f \in L^1_{\mathbb{R}}$, $\ell_s \in L_s$. As $(L^\infty_{\mathbb{R}})'$ is a Riesz space we may suppose that $\ell \geq 0$. Consider f and ℓ_s defined at the beginning of number 23. We have to show that ℓ_s is singular. By lemma 23 there exists $\varphi \in L^\infty_{\mathbb{R}}$ such that $\|\varphi\|_\infty \leq 1$ and

(1) $\langle \ell_s - \chi_T, \varphi \rangle \geq \|\ell_s\| + \|\chi_T\|_1 - \varepsilon$.

Thus $\int_T \varphi \, d\mu = < \chi_T, \varphi >$

$$\leq < \ell_s, \varphi > - \|\ell_s\| - \|\chi_T\|_1 + \varepsilon$$

$$\leq - \|\chi_T\|_1 + \varepsilon \text{ , hence}$$

(2) $\mu(\{t | \varphi(t) \geq 0\}) \leq \varepsilon$

We have also from (1)

$$< \ell_s, \varphi > \geq < \chi_T, \varphi > + \|\ell_s\| + \|\chi_T\|_1 - \varepsilon$$

$$\geq \|\ell_s\| - \varepsilon \text{ , hence (*)}$$

$$< \ell_s, \chi_{\{\varphi\geq 0\}} > = < \ell_s, \varphi + \chi_{\{\varphi\geq 0\}} (1-\varphi) - \chi_{\{\varphi<0\}} \varphi >$$

$$\geq \|\ell_s\| - \varepsilon + < \ell_s, \chi_{\{\varphi\geq 0\}} (1-\varphi) > - < \ell_s, \chi_{\{\varphi<0\}} \varphi >$$

$$\geq \|\ell_s\| - \varepsilon \text{ because } \ell_s \geq 0.$$

Therefore

(3) $\|\ell_s^{\{\varphi\geq 0\}}\| \geq \|\ell_s\| - \varepsilon.$

By lemma 11, (2) and (3) prove that ℓ_s is singular.

§5 - PROOF OF THE MAIN THEOREM WHEN μ IS BOUNDED

25 - We suppose that μ is bounded.

Lemma VIII.25. - Consider the set $\mathcal{V}$ of positive additive functions v on $\mathcal{C}$, such that $\mu(A) = 0 \Rightarrow v(A) = 0$, and the set $(L^\infty_{\mathbb{R}})'_+$ of positive continuous linear forms m on $L^\infty_{\mathbb{R}}$. Then there is a bijective correspondence between $\mathcal{V}$ and $(L^\infty_{\mathbb{R}})'_+$ such that the corresponding v and m satisfy the equality $v(A) = m(\chi_A)$.

Proof. If $m \in (L^\infty_{\mathbb{R}})'_+$ then v defined by $v(A) = m(\chi_A)$ belongs to $\mathcal{V}$. Conversely let $v \in \mathcal{V}$. Let ϕ be the subspace of $L^\infty_{\mathbb{R}}$ of equivalence classes of functions assuming a finite number of values. Then there exists one linear functional m_o on ϕ such that $m_o(\chi_A) = v(A)$. Moreover m_o

(*) $\{\varphi \geq 0\} = \{t | \varphi(t) \geq 0\}$

is positive and $|m_0(\varphi)| \leq v(T)\ \|\varphi\|_\infty$. Then the extension by continuity of m_0 to $L^\infty_{\mathbb{R}}$ has the required properties.

26 - Let $\ell \in (L^\infty_E)'$. For $A \in \mathcal{C}$, put $v(A) = \|\ell^A\|$ (ℓ^A is defined in prop. 10). Then v belongs to $\mathcal{V}$ (prop. 10). Let $\bar{v}$ denote the corresponding element of $(L^\infty_{\mathbb{R}})'_+$ (lemma 25). And let $\bar{v} = h + \bar{v}_s$ be the decomposition of $\bar{v}$ given by the main theorem : $h \in L^1_{\mathbb{R}}$ and $\bar{v}_s$ is singular.

<u>Lemma VIII.26.</u> - <u>We suppose that</u> $\bar{v}_s = 0$. <u>Then there exists</u> $f \in \mathcal{L}^1_{E'}[E]$ <u>such that</u> $\ell = \dot{f}$. <u>Moreover</u> $\|\dot{f}\|_1 = v(T)$.

a) The last part follows from lemma 3, because $v(T) = \|\ell\|$. Remark that

$$v(A) = \bar{v}(\chi_A) = \int_A h\ d\mu,$$ so that v is a measure.

For $A \in \mathcal{C}$, consider the map $x \mapsto \ell(\chi_A\ x)$. It is a continuous linear form on E which is defined by an element $\lambda(A)$ of E'. One has $\|\lambda(A)\| \leq v(A)$.

Thus, for every $x \in E$, $< \lambda(.),x >$ is a measure which has a bounded density

$$\frac{d < \lambda(.),x >}{dv}$$ with respect to v.

Let ρ be a lifting of $L^\infty_{\mathbb{R}}$. Then

$$x \mapsto \rho\left(\frac{d < \lambda(.),x >}{dv}\right)(t)$$ is a continuous linear form on E, and so defines an element $f_0(t)$ of E'. It is clear that f_0 is scalarly measurable and that $\|f_0(t)\| \leq 1$ for every t. Let us consider $f(t) = \frac{dv}{d\mu}(t)\ f_0(t)$. As $\frac{dv}{d\mu}$ is integrable, f belongs to $\mathcal{L}^1_{E'}[E]$.

We shall show that $\ell = \dot{f}$.

b) Remark that if (A_n) is a sequence of pairwise disjoint measurable sets which cover T, and if $\varphi \in L^\infty_{E'}$, then

$$\ell(\varphi) = \Sigma\ \ell(\chi_{A_n}\ \varphi)$$ because

$$\ell(\varphi) - \sum_{n=0}^{N} \ell(\chi_{A_n}) = \ell(\chi_{\bigcup_{n>N} A_n}\ \varphi)$$

and $\|\ell(\chi_{\bigcup_{n>N} A_n}\varphi)\| \le \|\varphi\|_\infty\ v(\bigcup_{n>N} A_n) \to 0$ when $N \to \infty$.

c) The subspace H of L^∞_E of the equivalence classes of functions assuming a countable set of values is dense in L^∞_E (because μ is bounded). If $\tilde{\varphi} \in H$ we may suppose that $\varphi = \Sigma\, \chi_{A_n} x_n$ where the A_n are pairwise disjoint and cover T, and $\|x_n\| \le \|\tilde{\varphi}\|_\infty$. One has

$$\begin{aligned}\ell(\varphi) &= \Sigma\, \ell(\chi_{A_n}\varphi) \quad \text{(by b))}.\\ &= \Sigma\, \ell(\chi_{A_n} x_n)\\ &= \Sigma < \lambda(A_n), x_n >\\ &= \Sigma \int_{A_n} < f_0(t), x_n > v(dt)\\ &= \int_T < f_0(t), \varphi(t) > v(dt)\\ &= \int_T < f(t), \varphi(t) > \mu(dt).\end{aligned}$$

By continuity that holds for every $\varphi \in L^\infty_E$.

27 - Proof of the main theorem

We return to the general case $\bar{v} = h + \bar{v}_s$. Let (A_n) be a decreasing sequence such that $\bar{v}_s$ is (A_n)-singular (prop. 9). We may suppose $A_0 = T$. Put $B_n = A_n - A_{n+1}$. Then the B_n are pairwise disjoint and $T - \bigcup B_n$ is negligible. If $A \in \mathcal{C}$, $A \subset B_n$, $\|\ell^A\| = v(A) = \bar{v}(\chi_A) = \int_A h\, d\mu$. Thus lemma 26 applies on B_n : there exists $f_n \in \mathcal{L}^1_{E'}[E, B_n]$ such that $\ell^{B_n}(\varphi) = \int_{B_n} < f_n(t), \varphi(t) > \mu(dt)$ and $\|\dot{f}_n\|_1 = v(B_n)$.

If we put $f(t) = \begin{cases} f_n(t) & \text{if } t \in B_n \\ 0 & \text{if } t \in T - \bigcup B_n \end{cases}$

Then f belongs to $\mathcal{L}^1_{E'}[E]$ because $\Sigma\, v(B_n) \le v(T)$.

It remains to show that $\ell_s = \ell - \dot{f}$ is singular. We shall show that ℓ_s is (A_n) singular. Let $\varphi \in L^\infty_E$ be zero on A_n.

Then $\varphi = \sum_{p=0}^{n-1} \chi_{B_p}\varphi$ and

$$\ell(\varphi) = \sum_{p=0}^{n-1} \ell^{B_p}(\varphi) = \int_T < f(t) \ \varphi(t) > \mu(dt).$$

So $\ell_s(\varphi) = 0.$

§6 - PROOF OF THE MAIN THEOREM

28 - Readers interested only in the case μ σ-finite will remark that a σ-finite-measureis equivalent to a bounded measure. So they will be not interested in the following proof.

Recall that $(T,\mathcal{C},\mu)$ has the direct sum property (see number 1).

Let $\ell \in (L_E^\infty)'$ and $k \in K$. For $\varphi \in L_E^\infty(T_k)$ let $\bar{\varphi}$ denote the function defined by

$$\bar{\varphi}(t) = \begin{cases} \varphi(t) & \text{if } t \in T_k \\ 0 & \text{if } t \in T - T_k. \end{cases}$$

Then $\ell_k(\varphi) = \ell(\bar{\varphi})$ defines an element ℓ_k of $L_E^\infty(T_k)'$.

By §5 $\ell_k = \dot{f}_k + \ell_k^s$ where $f_k \in \mathcal{L}^1_{E'}[E,T_k]$ and ℓ_k^s is singular.

Put

$$f(t) = \begin{cases} f_k(t) & \text{if } t \in T_k \\ 0 & \text{if } t \in T - \bigcup T_k \end{cases}$$

It is clear that f is scalarly measurable. And as $\Sigma \|\dot{f}\|_1 \leq \Sigma \|\ell_k\| \leq \|\ell\|$, f belongs to $\mathcal{L}^1_{E'}[E]$.

For every k let $(A_{k,n})$ be a sequence in $\mathcal{C} \cap \mathcal{P}(T_k)$ such that ℓ_k^s is $(A_{k,n})$ singular. Put $B_{k,n} = A_{k,n} \cup (T-T_k)$.

Then ess inf $B_{k,n} = \emptyset$. And if $\varphi \in L_E^\infty$ is zero on $B_{k,n}$ then

$$\begin{aligned} < \ell-\dot{f},\varphi > &= < \ell-\dot{f}, \chi_{T_k} \varphi > \\ &= \ell_k(\varphi) - < \dot{f}, \varphi|_{T_k} > \\ &= \ell_k^s(\varphi|_{T_k}) \\ &= 0. \end{aligned}$$

§7 - POLAR OF A CONVEX FUNCTIONAL ON L^{∞}.

29 - Let $(T, \mathcal{T}, \mu)$ be a σ-finite positive measure space, F a separable Banach space whose dual F' is strongly separable.

Let f be a convex normal integrand on $T \times F'$. (We suppose that the functions $f(t,.)$ are ℓ.s.c. for a topology compatible with the duality F,F'). By VII.2. we know that f^* defined by

$$f^*(t,x) = \sup\{< x',x > - f(t,x') | x' \in F'\}$$

is a convex normal integrand on $T \times F$.

We consider the spaces $L^1_{F'}$ and L^{∞}_F. Remark that as F' is separable $L^1_{F'} = L^1_{F'}[F]$. We consider the functionals defined in VII.7. :

$$I_f(u) = \int f(t,u(t))\mu(dt) \quad \text{for} \quad u \in L^1_{F'}$$

$$I_{f^*}(v) = \int f^*(t,v(t))\mu(dt) \quad \text{for} \quad v \in L^{\infty}_F.$$

Recall that theorem VII.7 says that if $\operatorname{dom} I_f$ and $\operatorname{dom} I_{f^*}$ are non empty then I_f and I_{f^*} are mutually polar. That means explicitly that

$$I_{f^*}(v) = \sup\{< u,v > - I_f(u) | u \in L^1_{F'}\}$$

$$I_f(u) = \sup\{< u,v > - I_{f^*}(v) | v \in L^{\infty}_F\}.$$

The following theorem (due to Rockafellar) gives a formula for $(I_{f^*})^*$ defined on $(L^{\infty}_F)'$ by

$$(I_{f^*})^*(\ell) = \sup\{< \ell,v > - I_{f^*}(v) | v \in L^{\infty}_F\}.$$

<u>Theorem VII.29.</u> - <u>Under the foregoing hypotheses including</u> $\operatorname{dom} I_f \neq \emptyset$ <u>and</u> $\operatorname{dom} I_{f^*} \neq \emptyset$, <u>one has</u>

$$(I_{f^*})^*(\ell) = I_f(h) + \delta^*(\ell_s | \operatorname{dom} I_{f^*})$$

<u>where</u> $\ell \in (L^{\infty}_F)'$ <u>and</u> $\ell = h + \ell_s$ $(h \in L^1_{F'},\ \ell_s \in L_s)$ <u>is its decomposition</u> (according to theorem 5).

<u>Proof</u>. First

$$\begin{aligned}(I_{f^*})^*(\ell) &= \sup\{< h,v > + \ell_s(v) - I_{f^*}(v) | v \in \operatorname{dom} I_{f^*}\}\\ &\leq \sup\{< h,v > - I_{f^*}(v) | v \in \operatorname{dom} I_{f^*}\} + \sup\{\ell_s(v') | v' \in \operatorname{dom} I_{f^*}\}\\ &= I_f(h) + \delta^*(\ell_s | \operatorname{dom} I_{f^*}).\end{aligned}$$

We shall now prove the converse inequality. By prop.9 (recall that μ is σ-finite) there exists a decreasing sequence (A_k) in $\mathcal{C}^\infty$ such that ℓ_s is (A_k)-singular. Let v_0 be any element of dom I_{f^*}, $\varepsilon > 0$ and $\alpha < I_f(h)$ $(\alpha \in \mathbb{R})$. Then for k large enough one has

$$\int_{A_k} [< h, v_0 > - f^*(.,v_0(.))] \mu > -\varepsilon \text{ , and}$$

$\int_{T-A_k} f(t,h(t))\mu(dt) > \alpha$ (recall that $f(t,h(t)) \geq < h(t), v_0(t) > - f^*(t,v_0(t))$ which is integrable). Then

$$\begin{aligned}(I_{f^*})^*(\ell) &= \sup \{\ell(v) - \int_{A_k} f^*(.,v(.))\mu - \int_{T-A_k} f^*(.,v(.))\mu \,|\, v \in L_F^\infty(T)\} \\ &= \sup\{< h,v > - \int_{T-A_k} f^*(.,v(.))\mu \,|\, v \in L_F^\infty(T - A_k)\} + \\ &+ \sup\{< h,v' > + \ell_s(v') - \int_{A_k} f^*(.,v'(.))\mu \,|\, v' \in L_F^\infty(A_k)\} \\ &= \int_{T-A_k} f(t,h(t))\mu(dt) + \sup\{\ell_s(v') + \int_{A_k} [<h,v'> - f^*(.,v'(.))]\mu \,|\, v' \in L_F^\infty\} \\ &\geq \int_{T-A_k} f(.,h(.))\mu + \ell_s(v_0) + \int_{A_k} [<h,v_0> - f^*(.,v_0(.))]\mu \\ &\geq \alpha + \ell_s(v_0) - \varepsilon .\end{aligned}$$

Hence for every $v_0 \in$ dom I_{f^*}

$(I_{f^*})^*(\ell) \geq I_f(h) + \ell_s(v_0)$ and then

$(I_{f^*})^*(\ell) \geq I_f(h) + \delta^*(\ell_s \,|\, \text{dom } I_{f^*})$.

30 - Under the foregoing hypotheses let us suppose that I_{f^*} is finite valued on the whole space L_F^∞. Then I_{f^*} is continuous with respect to the norm : indeed I_{f^*} is ℓ.s.c. (as supremum of continuous functions) and for every v_0 and $r > I_{f^*}(v_0)$ the subset of L_F^∞ defined as $\{v \,|\, I_{f^*}(v) \leq r\} - v_0$ is closed convex absorbing, hence a barrel, and so is a neighbourhood of 0 (we repeat here a well known argument). That proves that I_{f^*} is bounded above on a non empty open set, and so it is continuous by Bourbaki EVT ch II §2 prop. 21 p.60. But the following corollary says more : I_{f^*} is continuous for the Mackey topology $\tau(L_F^\infty, L_{F'}^1)$. Compare to theorems V-9 and V-11.

Corollary VIII.30. - Under the hypotheses of theorem 29, suppose that I_{f*} is finite valued on all L^{∞}_{F}. Then

a) I_{f*} is $\tau(L^{\infty}_{F}, L^{1}_{F'})$ continuous on L^{∞}_{F}

b) I_f is $\sigma(L^{1}_{F'}, L^{\infty}_{F})$ inf-compact for every slope

c) $(I_{f*})^*(\ell) = \begin{cases} I_f(h) & \text{if } \ell = h \in L^{1}_{F'} \\ +\infty & \text{otherwise.} \end{cases}$

Proof. c) results from theorem 29. Let us show b). For $v \in L^{\infty}_{F}$ and $\alpha \in \mathbb{R}$ the set

$\{\ell \in (L^{\infty}_{F})' \mid \ell(v) + (I_{f*})^*(\ell) \leq \alpha\}$ is $\sigma((L^{\infty}_{F})', L^{\infty}_{F})$ compact because I_{f*} is norm continuous on L^{∞}_{F} and from theorem I.12. But by c) (if we consider $L^{1}_{F'}$, as a subset of $(L^{\infty}_{F})'$) we have

$\{\ell \mid \ell(v) + (I_{f*})^*(\ell) \leq \alpha\} = \{h \in L^{1}_{F'} \mid \langle h, v \rangle + I_f(h) \leq \alpha\}$.

That proves b).

Finally b $\Rightarrow$ a thanks to theorem I.12.

§8 - CONDITIONAL EXPECTATION OF A RANDOM VECTOR.

31 - Prerequisites. Let $(\Omega, \mathcal{A}, P)$ denote a probability space and let $\mathcal{B}$ be a sub σ-field of $\mathcal{A}$. Then if $f \in \mathcal{L}^{1}_{\mathbb{R}}(\Omega, \mathcal{A}, P)$ there exists a function $g \in \mathcal{L}^{1}_{\mathbb{R}}(\Omega, \mathcal{B}, P)$ (we omit to write $P_{|\mathcal{B}}$!) such that

(CE) $\forall B \in \mathcal{B}$, $\int_B g \, dP = \int_B f dP$.

Such a function is called a conditional expectation of f. The terminology is justified by the following property : let $P^{\mathcal{B}}_{f}(.|.)$ be a $\mathcal{B}$-measurable regular conditional law of f, then the integral $\int_{\mathbb{R}} x \, P^{\mathcal{B}}_{f}(dx|\omega)$ converges a.e. and defines a conditional expectation of f. We shall not speak again of that property in the sequel. (For its interest in the study of conditional expections of multifunctions, see Daurès [5]).

It follows from property (CE) that if g and g' are two conditional expectations of f, then $g = g'$ a.e. And the equivalence class of g

depends only on the equivalence class of f. The corresponding map from $L^1_{\mathbb{R}}(\Omega,\mathcal{A},P)$ to $L^1_{\mathbb{R}}(\Omega,\mathcal{B},P)$ is denoted by $E^{\mathcal{B}}$. It is well known that $E^{\mathcal{B}}$ is linear, transforms positive functions into positive functions, and that $\|E^{\mathcal{B}}\| = 1$.

Remarks 1) Often the notation $E^{\mathcal{B}}(f)$ is used with $f \in \mathcal{L}^1_{\mathbb{R}}(\Omega,\mathcal{A},P)$ in place of $E^{\mathcal{B}}(\dot{f})$. And usually, $E^{\mathcal{B}}(f)$ is treated in formulas as a function.

2) Conditional expectation can be extended to quasi-integrable functions. A quasi-integrable function f is an $\mathcal{A}$-measurable function (with values in $\bar{\mathbb{R}}$) such that f^- is integrable. Then there exists a $\mathcal{B}$-measurable quasi-integrable function g(still denoted by $E^{\mathcal{B}}(f)$) such that

$$\forall B \in \mathcal{B}, \int_B g\, dP = \int_B f\, dP.$$

32 - Proposition VIII.32. - Let E be a separable Banach space and $\mathcal{B}$ a sub σ-field of $\mathcal{A}$. There exists a linear map $E^{\mathcal{B}} : L^1_E(\Omega,\mathcal{A},P) \to L^1_E(\Omega,\mathcal{B},P)$ of norm 1, such that (CE) $\forall B \in \mathcal{B}, \int_B E^{\mathcal{B}}(f)dP = \int_B f\, dP$.

Moreover $E^{\mathcal{B}}(f)$ is the unique element of $L^1_E(\Omega,\mathcal{B},P)$ satisfying (CE).

Proof. We first prove the uniqueness property. Suppose that g and g' satisfy (CE) and let (x'_n) be a dense sequence in E'.

Then it follows from (CE) that for every n,

$< x'_n, g >$ and $< x'_n, g' >$ are conditional expectations of $E^{\mathcal{B}}(< x'_n, f >)$.

Therefore $< x'_n, g > = < x'_n, g' >$ a.e. and so $g = g'$ a.e.

Consider now the following subspace H of L^1_E : H is the set of equivalence classes of functions $f = \sum_{i=1}^{n} \chi_{A_i} x_i$ ($n \in \mathbb{N}$ $A_i \in \mathcal{A}$, $x_i \in E$). If $f \in H$ let us define $E^{\mathcal{B}}(f)$ as $\sum E^{\mathcal{B}}(\chi_{A_i})x_i$. Clearly $E^{\mathcal{B}}(f)$ verifies (CE). As $E^{\mathcal{B}}(f)$ is unique, we obtain a linear map $E^{\mathcal{B}} : H \to L^1_E(\Omega,\mathcal{B},P)$. Let us show that $\|E^{\mathcal{B}}(f)\|_1 \le \|f\|_1$. We may suppose the A_i pairwise disjoint. First remark that

$$\|\Sigma E^{\mathcal{B}}(\chi_{A_i})(\omega)x_i\| \le \Sigma E^{\mathcal{B}}(\chi_{A_i})(\omega)\|x_i\| \text{ (because } E^{\mathcal{B}}(\chi_{A_i})(\omega) \ge 0)$$
$$= E^{\mathcal{B}}(\Sigma \chi_{A_i}\|x_i\|)(\omega)$$
$$= E^{\mathcal{B}}(\|\Sigma \chi_{A_i} x_i\|)(\omega) \quad \text{(because the } A_i \text{ are disjoint)}$$

(Remark : that inequality is a particular case of Jensen's inequality).

Then

$$\|E^{\mathcal{B}}(f)\|_1 = \int \|\Sigma E^{\mathcal{B}}(\chi_{A_i})(\omega)x_i\| P(d\omega)$$
$$\le \int E^{\mathcal{B}}(\|\Sigma \chi_{A_i} x_i\|)(\omega) \, P(d\omega)$$
$$= \int \|\Sigma \chi_{A_i} x_i\| \, dP$$
$$= \|f\|_1$$

Finally $E^{\mathcal{B}}$ can be extended to all $L^1_E(\Omega, \mathcal{A}, P)$ by continuity.

<u>Remark</u> Proposition 30 remains true for non separable Banach spaces. Indeed if $f : \Omega \to E$ is (strongly) measurable, there exists (see ch III. §0) a separable subspace E_o of E such that $f(\omega) \in E_o$ a.e. See Scalora [25].

33 - We give now a result in the special case that we shall need later.

<u>Proposition VIII.33. - Let</u> F <u>be a separable Banach space whose dual</u> F' <u>is strongly separable. Let</u> $\mathcal{B}$ <u>be a sub σ-field of</u> $\mathcal{A}$.
<u>Then the map</u> $E^{\mathcal{B}} : L^1_{F'}(\Omega, \mathcal{A}, P) \to L^1_{F'}(\Omega, \mathcal{B}, P)$ <u>and the embedding map</u> $\text{inj} : L^\infty_F(\Omega, \mathcal{B}, P) \to L^\infty_F(\Omega, \mathcal{A}, P)$ <u>are mutually transpose</u>.

<u>Proof</u> Let $u \in L^1_{F'}(\Omega, \mathcal{A}, P)$ and $v \in L^\infty_F(\Omega, \mathcal{B}, P)$.

We have to prove that

$$< E^{\mathcal{B}}(u), v > = < u, v >.$$

If $u = \chi_A x'$ $(A \in \mathcal{A},\ x' \in F')$ that is true because

$E^{\mathcal{B}}(u) = E^{\mathcal{B}}(\chi_A)x'$ and

$$< E^{\mathcal{B}}(u), v > = \int E^{\mathcal{B}}(\chi_A)(\omega) < x', v(\omega) > P(d\omega)$$
$$= \int E^{\mathcal{B}}[\chi_A < x', v >] dP \quad (\text{because } < x', v > \text{ is } \mathcal{B} \text{ measurable})$$
$$= \int \chi_A < x', v > dP$$
$$= < u, v >$$

Then the formula is valid if $u = \sum_{i=1}^{n} \chi_{A_i} x'_i$ by linearity, and for every $u \in L^1_E(\Omega, \mathcal{A}, P)$ by continuity.

§9 - CONDITIONAL EXPECTATIONS OF INTEGRANDS AND RANDOM SETS.

34 - Let F be separable Banach space, whose dual F' is strongly separable. Let $(\Omega, \mathcal{A}, P)$ be a probability space, $\mathcal{B}$ a sub σ-field of $\mathcal{A}$. For technical reasons we suppose $\mathcal{A}$ and $\mathcal{B}$ complete.

Definition. We shall say that Γ is a $\mathcal{A}$-random (resp $\mathcal{B}$-random) closed convex set in F' if Γ is a $\mathcal{A}$-measurable (resp. $\mathcal{B}$-measurable multifunction) (see theorem I-30) with $\sigma(F',F)$ closed convex values.

In the sequel we denote by $B_{F'}$ the unit ball of F'. Let Γ be a $\mathcal{A}$-random closed convex set in F'. We suppose that there exists $\alpha \in \mathcal{L}^1_+(\Omega, \mathcal{A}, P)$ such that

$$\Gamma(\omega) \subset \alpha(\omega)\, B_{F'} \text{ a.e.}$$

Definition. A $\mathcal{B}$-random closed convex set in F', Σ, such that there exists $\beta \in \mathcal{L}^1_+(\Omega, \mathcal{B}, P)$ with

$$\Sigma(\omega) \subset \beta(\omega)\, B_{F'} \text{ a.e.}$$

is called a conditional expectation of Γ if

(CE) $\forall\, B \in \mathcal{B}$, $\int_B \Sigma\, dP = \int_B \Gamma\, dP$. (we use the notation of V.14.)

By theorem V.14 and its remark (F' with $\sigma(F',F)$ is Suslin and its dual is F), condition (CE) is equivalent to

$$\forall\, B \in \mathcal{B},\ \forall\, x \in F,\ \int_B \delta^*(x|\Sigma(\omega))P(d\omega) = \int_B \delta^*(x|\Gamma(\omega))P(d\omega)$$

which is equivalent to

$$\forall\, x \in F,\ \delta^*(x|\Sigma(\omega)) = E^{\mathcal{B}}[\delta^*(x|\Gamma(.))](\omega) \text{ a.e.}$$

Theorem VIII.34. - Under the foregoing hypotheses there exists a unique (for equality a.e.) conditional expectation of Γ, Σ. Moreover Σ has the properties

a) $\Sigma(\omega) \subset E^{\mathcal{B}}(\alpha)B_{F'}$ a.e.

b) $\forall\, v \in L^{\infty}_{F}(\Omega,\mathcal{B},P)$, $\int \delta^*(v(\omega)|\Sigma(\omega))P(d\omega) = \int \delta^*(v(\omega)|\Gamma(\omega))P(d\omega)$

c) $S_{\Sigma}(\mathcal{B}) = E^{\mathcal{B}}(S_{\Gamma}(\mathcal{A}))$ (the notation S_{Γ} is that of V.13.

We specify here the σ-field)

We shall denote $E^{\mathcal{B}}(\Gamma)$ by Σ.

Proof. 1) To prove the existence of Σ we apply theorem V.16.

We take $\Lambda = L^1_{\mathbb{R}}(\Omega,\mathcal{B},P)$ and $\Lambda^* = L^{\infty}_{\mathbb{R}}(\Omega,\mathcal{B},P)$.

We put $M(f) = \int f\,\Gamma\,dP$ for $f \in \Lambda^*$.

By theorem V.16 there exists a $\mathcal{B}$ measurable multifunction Σ (here $\mathcal{B}$ replaces $\mathcal{A}$ and Σ replaces Γ) such that $\forall f \in \Lambda^*$, $M(f) = \int f\,\Sigma\,dP$.

Taking $f = \chi_B$ $(B \in \mathcal{B})$ we obtain condition (CE).

2) We prove now the uniqueness. Let Σ_1 and Σ_2 be two conditional expectations. Then for every $x \in F$, $\delta^*(x|\Sigma_1(.)) = \delta^*(x|\Sigma_2(.))$ a.e. because these functions are conditional expectations of $\delta^*(x|\Gamma(.))$.

By definition $\Sigma_1(\omega)$ and $\Sigma_2(\omega)$ are convex $\sigma(F',F)$ compact. So by prop. III.35. $\Sigma_1 = \Sigma_2$ a.e.

3) Now we prove a). For $x \in F$

$$\delta^*(x|\Sigma(\omega)) = E^{\mathcal{B}}(\delta^*(x|\Gamma(.)))(\omega) \le E^{\mathcal{B}}(\|x\|\alpha)(\omega)$$
$$= \|x\|\; E^{\mathcal{B}}(\alpha)(\omega) = E^{\mathcal{B}}(\alpha)(\omega)\,\delta^*(x|B_{F'})$$

So again by prop III.35 we have $\Sigma(\omega) \subset E^{\mathcal{B}}(\alpha)(\omega)\; B_{F'}$ a.e.

4) Now we prove b). First suppose that

$v = \sum_{i=1}^{n} \chi_{B_i}\; x_i$ $(B_i \in \mathcal{B}$, the B_i disjoint, $x_i \in F)$.

Then the formula

$$\int \delta^*(v(\omega)|\Sigma(\omega))P\,(d\omega) = \int \delta^*(v(\omega)|\Gamma(\omega))P(d\omega)$$

results from (CE). Remark that the space of all $v = \sum \chi_{B_i}\; x_i$ is $\tau(L^{\infty}_{F}, L^{1}_{F'})$ dense in L^{∞}_{F} because its orthogonal is 0.

But by corollary 30 the functionals in each member of the formula are $\tau(L^{\infty}_{F'}, L^{1}_{F'})$ continuous.

5) It remains to prove c). First we prove that $E^{\mathcal{B}}(S_{\Gamma}(\mathcal{A})) \subset S_{\Sigma}(\mathcal{B})$.

Let $u \in S_{\Gamma}(\mathcal{A})$, and $x \in F$. Then

$$< E^{\mathcal{B}}u,x > \leq E^{\mathcal{B}}[\delta^*(x|\Gamma(.))] \text{ a.e.}$$

$$= \delta^*(x|\Sigma(.)) \text{ a.e.}$$

So again by III.35 $E^{\mathcal{B}}u \in S_{\Sigma}(\mathcal{B})$.

We prove now the opposite inclusion. First remark that $S_{\Gamma}(\mathcal{A})$ is $\sigma(L^{1}_{F'}, L^{\infty}_{F})$ compact. Indeed let $\Gamma'(\omega) = (1+\alpha(\omega))^{-1}\Gamma(\omega)$; one has $\Gamma'(\omega) \subset B_{F'}$. Then S_{Γ} is the range of the map $A : u \to (1+\alpha)u$ from $S_{\Gamma'}$ to $L^{1}_{F'}$. Let $v \in L^{\infty}_{F}$. One has $<A(u),v> = <(1+\alpha)u,v> = <u,(1+\alpha)v>$. As $(1+\alpha)v \in L^{1}_{F}$ that proves that A is $\sigma(L^{\infty}_{F'},L^{1}_{F})$, $\sigma(L^{1}_{F'},L^{\infty}_{F})$ continuous. By theorem V.1 $S_{\Gamma'}$ is $\sigma(L^{\infty}_{F'},L^{1}_{F})$ compact. Therefore $A(S_{\Gamma'}) = S_{\Gamma}$ is $\sigma(L^{1}_{F'}, L^{\infty}_{F})$ compact. Now thanks to prop. 33 $E^{\mathcal{B}}$ is continuous for the topologies $\sigma(L^{1}_{F'}, L^{\infty}_{F})$ so $E^{\mathcal{B}}(S_{\Gamma}(\mathcal{A}))$ is $\sigma(L^{1}_{F'}(\mathcal{B}), L^{\infty}_{F}(\mathcal{B}))$ compact. Then it suffices to prove that $\forall v \in L^{\infty}_{F}(\mathcal{B})$, $\delta^*(v|E^{\mathcal{B}}(S_{\Gamma}(\mathcal{A}))) \geq \delta^*(v|S_{\Sigma}(\mathcal{B}))$.

Let $u \in S_{\Gamma}(\mathcal{A})$ such that $< u(\omega),v(\omega) > = \delta^*(v(\omega)|\Gamma(\omega))$ (such a selection exists because $\Gamma(\omega)$ is $\sigma(F',F)$ compact and thanks to theorem III.22 or III.30). One has

$$\delta^*(v|E^{\mathcal{B}}(S_{\Gamma}(\mathcal{A}))) \geq < E^{\mathcal{B}}(u),v >$$

$$= < u,v > \text{ (by prop. 33)}$$

$$= \int \delta^*(v(\omega)|\Gamma(\omega))P(d\omega)$$

$$= \int \delta^*(v(\omega)|\Sigma(\omega))P(d\omega) \quad \text{(by b))}$$

$$\geq < u_1,v > \text{ for any } u_1 \in S_{\Sigma}(\mathcal{B}).$$

35 - Notation. Let Γ be a $\mathcal{A}$-random set in F'. We denote by $\mathcal{L}^{1}_{\Gamma}(\Omega, \mathcal{A}, P)$ the set $\{u \in \mathcal{L}^{1}_{F'}(\Omega,\mathcal{A},P) | u(\omega) \in \Gamma(\omega) \text{ a.e.}\}$ and by $L^{1}_{\Gamma}(\Omega,\mathcal{A},P)$ the quotient.

Theorem VIII.35. Let Γ be a $\mathcal{A}$-random closed convex set in F' which admits an integrable selection u_o. For every $n \in \mathbb{N}$ let

$\Gamma_n(\omega) = \Gamma(\omega) \cap (u_o(\omega) + n B_{F'})$ and

$\Sigma(\omega) = \overline{\cup E^{\mathcal{B}}(\Gamma_n)(\omega)}$. Then $\Sigma(\omega)$, which is a.e. convex, is a $\mathcal{B}$-random closed convex set. Moreover

a) $\forall v \in L_F^\infty$, $\int \delta^*(v(\omega)|\Sigma(\omega))P(d\omega) = \int \delta^*(v(\omega)|\Gamma(\omega))P(d\omega)$

In particular, taking $v = \chi_B x$ ($B \in \mathcal{B}$, $x \in F$), one obtains the property of conditional expectation.

b) Σ is the unique (for the equality a.e.) $\mathcal{B}$-random closed convex set with property a).

c) Σ is the smallest (for inclusion a.e.) of the $\mathcal{B}$-random closed convex sets Θ such that $L^1_\Theta(\mathcal{B}) \supset E^{\mathcal{B}}(L^1_\Gamma(\mathcal{A}))$.

We shall denote $\widehat{E^{\mathcal{B}}(\Gamma)}$ by Σ and say that Σ is the conditional expectation of Γ. Moreover $E^{\mathcal{B}}$ is increasing : that is $\Gamma_1(\omega) \subset \Gamma_2(\omega)$ a.e. implies $E^{\mathcal{B}}\Gamma_1 \subset E^{\mathcal{B}}\Gamma_2$ a.e.

Proof. 1) The sequence $\Gamma_n(\omega)$ is increasing. So by III.35, $E^{\mathcal{B}}(\Gamma_n)(\omega)$ is a.e. increasing, and so $\Sigma(\omega)$ is a.e. convex. Moreover Σ has a measurable graph (with respect to $\mathcal{B}$).

2) Property a) is a consequence of theorem 34 and the monotone convergence theorem. Indeed one has

$\forall n, \int \delta^*(v(\omega)|\Gamma_n(\omega))P(d\omega) = \int \delta^*(v(\omega)|E^{\mathcal{B}}(\Gamma_n)(\omega))P(d\omega),$

$< u_o, v > \leq \delta^*(v|\Gamma_n) \nearrow \delta^*(v|\Gamma),$

$< E^{\mathcal{B}}u_o, v > \leq \delta^*(v|E^{\mathcal{B}}\Gamma_n) \nearrow \delta^*(v|\Sigma).$

3) We prove now the last part of the statement : if $\Gamma_1 \subset \Gamma_2$ a.e. and if Σ_1 and Σ_2 satisfy a) then $\Sigma_1 \subset \Sigma_2$ a.e. That will prove also b). Suppose that the set $A = \{\omega | \Sigma_1(\omega) \not\subset \Sigma_2(\omega)\}$ is non negligible. Remark that A belongs to $\mathcal{B}$: indeed, if $G(\Sigma_i)$ denotes the graph of Σ_i,

$A = pr_\Omega[G(\Sigma_1) - G(\Sigma_2)]$, and the completeness of $\mathcal{B}$ and theorem III.23 prove $A \in \mathcal{B}$. The multifunction K defined on A by $K(\omega) = \{x \in F \mid \delta^*(x|\Sigma_1(\omega)) > \delta^*(x|\Sigma_2(\omega))\}$ has non empty values and a measurable graph. (the functions $(\omega,x) \mapsto \delta^*(x|\Sigma_i(\omega))$ are measurable by lemma VII.1). Therefore by theorem III.22 K has a measurable selection v_o. There exists a non negligible subset B of A on which v_o is bounded. Taking $v(\omega) = \begin{cases} v_o(\omega) & \text{if } \omega \in B \\ 0 & \text{if } \omega \in \Omega - B \end{cases}$

one obtains $\int \delta^*(v|\Sigma_1)dP > \int \delta^*(v|\Sigma_2)dP$.

4) We prove now c). First $L^1_\Sigma(\mathcal{B}) \supset E^{\mathcal{B}}(L^1_\Gamma(\mathcal{Q}))$.

Indeed if $u \in L^1_\Gamma(\mathcal{Q})$, $E^{\mathcal{B}}(u)$ belongs to $L^1_\Sigma(\mathcal{B})$ by the increasing property of $E^{\mathcal{B}}$ proved in 3).

Finally we have to prove that if $L^1_\Theta(\mathcal{B}) \supset E^{\mathcal{B}}(L^1_\Gamma(\mathcal{Q}))$ then $\Theta \supset \Sigma$ a.e. For every n $L^1_\Theta(\mathcal{B}) \supset E^{\mathcal{B}}(L^1_{\Gamma_n}(\mathcal{Q})) = L^1_{\Sigma_n}(\mathcal{B})$ (with $\Sigma_n = E^{\mathcal{B}}\Gamma_n$). But that implies $\Theta(\omega) \supset \Sigma_n(\omega)$ a.e. : indeed if $\{\omega|\Sigma_n(\omega) \not\subset \Theta(\omega)\}$ is non negligible one can exhibit by standard arguments a selection belonging to $L^1_{\Sigma_n}(\mathcal{B}) - L^1_\Theta(\mathcal{B})$. Hence $\Theta(\omega) \supset \Sigma(\omega)$ a.e.

<u>Remark</u> The functions $\delta^*(x'|\Sigma(.))$ and $\delta^*(x'|\Gamma(.))$ are quasi integrable and verify

$$\delta^*(x'|\Sigma(\omega)) = E^{\mathcal{B}}(\delta^*(x'|\Gamma(.)|)(\omega).$$

That property, which is weaker than a), does not permit us to obtain uniqueness, because we cannot apply prop. III.35.

36 - We extend now theorem 35 to integrands. See Bismut [1] for another method and for many refinements.

<u>Theorem VIII.36.</u> - a) <u>Let</u> $f : \Omega \times F' \to \bar{\mathbb{R}}$ <u>be a</u> $\mathcal{Q} \otimes \mathcal{B}(F')$ <u>measurable normal convex integrand and</u> f^* <u>its polar</u>. <u>We suppose that there exists</u>

$u_o \in \mathcal{L}^1_{F'}(\Omega, \mathcal{A}, P)$ <u>such that</u> $f^+(\omega, u_o(\omega))$ <u>is integrable. Then there exists</u> a $\mathcal{B} \otimes \mathcal{B}(F')$ <u>measurable normal convex integrand</u> g <u>such that</u> $\forall v \in L^\infty_F(\Omega, \mathcal{B}, P)$, <u>the functions</u> $f^*(.,v(.))$ <u>and</u> $g^*(.,v(.))$ <u>are quasi-integrable, and</u> $\forall B \in \mathcal{B}$ $\int_B f^*(.,v(.))dP = \int_B g^*(.,v(.))dP$.
<u>Moreover such a</u> g <u>is unique for the equality a.e. and the mapping</u> $f \mapsto g$ <u>is increasing</u>.

b) <u>Let</u> Γ <u>be a</u> $\mathcal{A}$<u>-random closed convex set such that</u> $L^1_\Gamma(\mathcal{A}) \neq \emptyset$, <u>let</u> $f(\omega, x') = \delta(x' | \Gamma(\omega))$ <u>and</u> Σ <u>be the conditional expectation of</u> Γ <u>described in theorem 35. Then if</u> g <u>is the integrand given by</u> a) <u>one has</u>

$$g(\omega, .) = \delta(. | \Sigma(\omega)) \text{ for almost every } \omega.$$

<u>Proof</u> a) Let $\Gamma(\omega) = \text{epi } f(\omega,.)$. It is a $\mathcal{A}$-random closed convex set in $F' \times \mathbb{R}$. And Γ admits an integrable selection : $\omega \to (u_o(\omega), f^+(\omega, u_o(\omega)))$. Let Σ be the conditional expectation of Γ given by theorem 35. We prove now that a.e. $\Sigma(\omega)$ is an epigraph. Put $r(\omega) = f^+(\omega, u_o(\omega))$. Then for every $n \in \mathbb{N}$ $(u_o(\omega), r(\omega) + n) \in \Gamma(\omega)$, and

$E^{\mathcal{B}}(u_o, r+n) = E^{\mathcal{B}}(u_o, r) + (0,n)$ is a selection of Σ.

If N is a negligible such that $\forall \omega \in \Omega - N$, $\forall n \in \mathbb{N}$

$$E^{\mathcal{B}}(u_o, r)(\omega) + (0,n) \in \Sigma(\omega)$$

we have $\forall \omega \in \Omega - N$, $\Sigma(\omega) \supset E^{\mathcal{B}}(u_o, r)(\omega) + \{0\} \times [0, \infty[$ and so $\Sigma(\omega)$ is an epigraph. Let g be the corresponding integrand. More precisely we can put

$$g(\omega, x') = \begin{cases} \inf \{\lambda \in \mathbb{R} | (x', \lambda) \in \Sigma(\omega)\} & \text{if } \omega \in \Omega - N \\ 0 \text{ if } \omega \in N. \end{cases}$$

As $f^*(\omega, v(\omega)) = \delta^*((v(\omega), -1) | \Gamma(\omega))$ and

$g^*(\omega, v(\omega)) = \delta^*((v(\omega), -1 | \Sigma(\omega))$, these functions are quasi-integrable, and it follows from theorem 35 that they have the same integral on every $B \in \mathcal{B}$.

Finally the uniqueness and the increasing property can be proved as in theorem 35. We sketch the proof. Let $f_1 \leq f_2$ and g_1, g_2 be the corresponding integrands. Suppose that $A = \{\omega | g_1(\omega,.) \nleq g_2(\omega,.)\}$ is non negligible. Then A belongs to $\mathcal{B}$ and $A = \{\omega | g_1^*(\omega,.) \ngeq g_2^*(\omega,.)\}$. Then the multifunction K defined on A by $K(\omega) = \{x \in F | g_1^*(\omega,x) < g_2^*(\omega,x)\}$ has non empty values.
It admits a measurable selection v (by III.22). There exists a non negligible set $B \in \mathcal{B}$, on which v is bounded. One obtains the following contradiction

$$\int_B f_1^*(.,v(.))dP = \int_B g_1^*(.,v(.))dP$$
$$< \int_B g_2^*(.,v(.))dP$$
$$= \int_B f_2^*(.,v(.))dP.$$

b) Let $u_0 \in \mathcal{L}^1_\Gamma(\mathcal{Q})$. Then u_0 satisfies the hypothesis of a).
For every $v \in L^\infty_F(\mathcal{B})$ and $B \in \mathcal{B}$

$$\int_B g^*(\omega,v(\omega))P(d\omega) = \int_B f^*(\omega,v(\omega))P(d\omega) \quad \text{and}$$

$$\int_B \delta^*(v(\omega)|\Sigma(\omega))P(d\omega) = \int_B \delta^*(v(\omega)|\Gamma(\omega))P(d\omega)$$
$$= \int_B f^*(\omega,v(\omega))P(d\omega).$$

The uniqueness property proved in a) entails $g^* = \delta^*(.|\Sigma(.))$.

<u>Remark</u>. The integrand g^* can be called the conditional expectation of f^*. It has the property $\forall x \in F$, $g^*(\omega,x) = E^{\mathcal{B}}[f^*(.,x)](\omega)$. On the other hand g is a kind of conditional infimum convolution of the functions $f(\omega,.)$. In the particular case $\mathcal{B} = \{\emptyset,\Omega\}$, where $\mathcal{B}$-measurable functions are constant, one has $g^*(x) = \int_\Omega f^*(\omega,x)P(d\omega)$
and its polar g is the greatest $\ell.s.c.$ function less than
$x' \mapsto \inf\{\int_\Omega f(\omega,u(\omega))P(d\omega) | u \in L^1_{F'}(\mathcal{Q}), \int u dP = x'\}$.
That will be proved in theorem 40 below.

37 - We denote by rest: $(L_F^\infty(\mathcal{A}))' \to (L_F^\infty(\mathcal{B}))'$ the restriction map defined (for $\ell \in (L_F^\infty(\mathcal{A}))'$, $v \in L_F^\infty(\mathcal{B})$) by $[\text{rest}(\ell)](v) = \ell(v)$. It is obvious that rest is the transpose of the embedding map inj : $L_F^\infty(\mathcal{B}) \to L_F^\infty(\mathcal{A})$. Moreover by prop. 33 if $u \in L_{F'}^1(\mathcal{A})$, $\text{rest}(u) = E^{\mathcal{B}}(u)$.

We keep the hypotheses of th. 36. We have $I_{g^*} = I_{f^*} \circ \text{inj}$. Let $(I_{g^*})^*$ and $(I_{f^*})^*$ be the polar considered on the dual spaces $(L_F^\infty(\mathcal{B}))'$ and $(L_F^\infty(\mathcal{A}))'$. Suppose that I_{f^*} is finite and norm continuous at a point $v_0 \in L_F^\infty(\mathcal{B})$. Then by theorem I.29

$\partial' I_{g^*}(v) = \text{rest}[\partial' I_{f^*}(v)]$ (we denote by ∂' the subdifferential calculated in $(L_F^\infty)'$).

__Theorem VIII.37.__ - __Let__ f __and__ g __be as in theorem__ 36. __Suppose that__ I_{f^*} __is finite and norm continuous at__ $v_0 \in L_F^\infty(\mathcal{B})$. __Then for every__ $v \in L_F^\infty(\mathcal{B})$

$$\partial I_{g^*}(v) = E^{\mathcal{B}}(\partial I_{f^*}(v)) + \text{As}[\partial I_{g^*}(v)].$$

(__here__ ∂ __denotes the subdifferential calculated in__ $L_{F'}^1$ __and__ As __denotes the asymptotic cone defined in__ I.7.)

__Proof__. First let $u \in \partial I_{f^*}(v)$. Let $v' \in L_F^\infty(\mathcal{B})$. Then

$I_{f^*}(v+v') \geq I_{f^*}(v) + \langle u, v' \rangle$, hence

$I_{g^*}(v+v') \geq I_{g^*}(v) + \langle E^{\mathcal{B}}(u), v' \rangle$ (by prop. 33).

That proves $E^{\mathcal{B}}(\partial I_{f^*}(v)) \subset \partial I_{g^*}(v)$. So the inclusion

$\partial I_{g^*}(v) \supset E^{\mathcal{B}}(\partial I_{f^*}(v)) + \text{As}[\partial I_{g^*}(v)]$ is true.

We prove now the converse inequality. Let $u \in \partial I_{g^*}(v)$. Remark that

$\partial I_{g^*}(v) \neq \emptyset \Rightarrow I_{g^*}(v) < \infty \Leftrightarrow I_{f^*}(v) < \infty$. By the foregoing remarks $u = \text{rest}(\ell)$ with $\ell \in \partial' I_{f^*}(v)$. By proposition I.26.

$\ell \in \partial' I_{f^*}(v) \Leftrightarrow I_{f^*}(v) + (I_{f^*})^*(\ell) = \ell(v)$.

Let $\ell = u_1 + \ell_s$ be the decomposition of ℓ. By theorem 29

$\ell \in \partial' I_{f^*}(v) \Leftrightarrow I_{f^*}(v) + I_f(u_1) + \delta^*(\ell_s \mid \text{dom } I_{f^*}) = \langle u_1, v \rangle + \ell_s(v)$.

But we have the inequalities

$\delta^*(\ell_s | \text{dom } I_{f^*}) \geq \ell_s(v)$ (because $v \in \text{dom } I_{f^*}$)

$I_{f^*}(v) + I_f(u_1) \geq \langle u_1, v \rangle$ (prop. I.26)

So

$$\ell \in \partial' I_{f^*}(v) \Leftrightarrow \begin{cases} I_{f^*}(v) + I_f(u_1) = \langle u_1, v \rangle \\ \text{and } \delta^*(\ell_s | \text{dom } I_{f^*}) = \ell_s(v) \end{cases}$$

$$\Leftrightarrow (I) \begin{cases} u_1 \in \partial I_{f^*}(v) \\ \text{and } \delta^*(\ell_s | \text{dom } I_{f^*}) = \ell_s(v) \end{cases}$$

Thus $u = \text{rest}(\ell) = E^{\mathcal{B}}(u_1) + \text{rest}(\ell_s)$, with $u_1 \in \partial I_{f^*}(v)$.

It remains to prove that $\text{rest}(\ell_s) \in \text{As}[\partial I_{g^*}(v)]$ or equivalently that

$\forall r \geq 0,\ E^{\mathcal{B}}(u_1) + r \ \text{rest}(\ell_s) \in \partial I_{g^*}(v)$.

But u_1 and $r\ \ell_s$ satisfy relations (I) above. So

$u_1 + r\ \ell_s \in \partial' I_{f^*}(v)$, and one has

a) $\text{rest}(u_1 + r\ell_s) \in \partial' I_{g^*}(v)$

b) $\text{rest}(u_1 + r\ell_s) = E^{\mathcal{B}}(u_1) + r \ \text{rest}(\ell_s) \in L^1_{F'}(\mathcal{B})$.

Thus $E^{\mathcal{B}}(u_1) + r \ \text{rest}(\ell_s) \in \partial' I_{g^*}(v) \cap L^1_{F'}(\mathcal{B}) = \partial I_{g^*}(v)$

38 - Corollary VIII.38. - Let Γ be a $\mathcal{A}$-random closed convex set in F', such that $L^1_\Gamma(\mathcal{A}) \neq \emptyset$. Suppose that there exists a $\sigma(F', F)$ closed convex locally compact set which contains no line, C, such that $\forall \omega$, $\Gamma(\omega)) \subset C$. Let $\Sigma = E^{\mathcal{B}}\Gamma$ be the conditional expectation given by the theorem 35. Then

a) $\Sigma(\omega) \subset C$ a.e.

b) $L^1_\Sigma(\mathcal{B}) = E^{\mathcal{B}}(L^1_\Gamma(\mathcal{A})) + L^1_{\text{As}(\Sigma)}(\mathcal{B})$.

Proof a) The inclusion $\Sigma(\omega) \subset C$ results from the increasing property of $E^{\mathcal{B}}$, which has been proved in theorem 35.

But here we can also refer to prop III.35.

b) Let $f(\omega, x') = \delta(x' | \Gamma(\omega))$. Remark that the hypotheses of theorem 36 are verified. We shall use the fact that $\partial I_{f^*}(0) = L^1_\Gamma(\mathcal{A})$: indeed

$u \in \partial I_{f*}(0) \Leftrightarrow \forall v \in L^\infty_F(\mathcal{A}),\ I_{f*}(0) + < u,v > \leq I_{f*}(v)$

$\Leftrightarrow \forall v \in L^\infty_F(\mathcal{A}),\ \int < u(\omega),v(\omega) > P(d\omega) \leq \int \delta^*(v(\omega)|\Gamma(\omega))P(d\omega)$

$\Leftrightarrow u \in L^1_\Gamma(\mathcal{A})$ (by standard arguments)

Similarly $\partial I_{g*}(0) = L^1_\Sigma(\mathcal{B})$.

Let us show now that the hypotheses of theorem 37 are verified. By corollary I.15. there exists $x_0 \in F$ such that $\delta^*(.|C)$ is finite and norm continuous at x_0 (on F the norm topology is the Mackey topology). So there exists $\epsilon > 0$ and $M \in \mathbb{R}$ such that $\|x - x_0\| \leq \epsilon \Rightarrow \delta^*(x|C) \leq M$. Hence if $v \in L^\infty_F(\mathcal{A})$ is such that $\|v - \chi_\Omega x_0\|_\infty \leq \epsilon$ one has

$$I_{f*}(v) = \int \delta^*(v(\omega)|\Gamma(\omega))P(d\omega)$$

$$\leq \int \delta^*(v(\omega)|C)P(d\omega)$$

$$\leq M.$$

On the other hand $I_{f*}(v) > -\infty$ because $L^1_\Gamma(\mathcal{A}) \neq \emptyset$. That proves that I_{f*} is finite and norm continuous at $\chi_\Omega x_0$ which belongs to $L^\infty_F(\mathcal{B})$.

Then the formula of th.37 holds and it can be written

$$L^1_\Sigma(\mathcal{B}) = E^{\mathcal{B}}(L^1_\Gamma(\mathcal{A})) + As[L^1_\Sigma(\mathcal{B})].$$

It remains to prove $As[L^1_\Sigma(\mathcal{B})] = L^1_{As(\Sigma)}(\mathcal{B})$.

Let $u_0 \in L^1_\Sigma(\mathcal{B})$. First if $u \in As[L^1_\Sigma(\mathcal{B})]$, one has for every $n \in \mathbb{N}$ $u_0(\omega) + n\, u(\omega) \in \Sigma(\omega)$ if $\omega \notin N_n$ (N_n negligible). So if $\omega \notin \bigcup_n N_n$, $u(\omega) \in As(\Sigma(\omega))$. Conversely if $u \in L^1_{As(\Sigma)}(\mathcal{B})$, then for any $r \geq 0$, $u_0(\omega) + ru(\omega) \in \Sigma(\omega)$ a.e. and $u \in As(L^1_\Sigma(\mathcal{B}))$.

Remark. It can be easily proved that $As(\Sigma)$ is a $\mathcal{B}$-random set (recall prop. I.7).

39 - Corollary VIII.39. - *Let $(T,\mathcal{T},\mu)$ be a σ-finite measure space. Let Γ be a measurable multifunction with $\sigma(F',F)$ closed convex values. Suppose $\Gamma(t) \subset \alpha(t)C$ a.e. where $\alpha \in \mathcal{L}^1_+(\mathcal{A})$ and C is a closed convex*

<u>locally compact set which contains no line and that</u> Γ <u>admits an integrable selection</u>.

<u>Then, if we let</u>

$$w\int \Gamma\, d\mu = \{x' \in F' \mid \forall\, x \in F,\ < x',x > \le \int \delta^*(x|\Gamma(t))\mu(dt)\}$$

<u>and</u> $\int \Gamma\, d\mu = \{\int u\, d\mu \mid u \in L^1_\Gamma\}$, <u>then one has</u>

$$w \int \Gamma\, d\mu = \int \Gamma\, d\mu + As[w \int \Gamma\, d\mu]$$

<u>and</u> $w \int \Gamma\, d\mu = \overline{\int \Gamma\, d\mu}$.

<u>Proof.</u> 1) As μ is σ-finite there exists an integrable function $\beta : T \to]0,\infty[$ such that $\beta(t) \ge \alpha(t)$. We may replace C by $\overline{co}(C \cup \{0\})$ which is still locally compact (that follows from corollary I.15 and the fact that $\delta^*(x|\overline{co}(C \cup \{0\})) = \sup(\delta^*(x|C,0))$.

Then $\Gamma'(t) = \frac{1}{\beta(t)}\ \Gamma(t) \subset \frac{\alpha(t)}{\beta(t)}\, C \subset C$ (because $0 \in C$ and $\frac{\alpha(t)}{\beta(t)} \le 1$)

If we replace Γ by Γ' we have to integrate with respect to $\beta\mu$ which is a bounded measure. Finally we may suppose that $\alpha = 1$ and that μ is a probability. As integrals are conditional expectations with respect to $\{\emptyset,T\}$ we shall apply corollary 38.

We have

$\Sigma = E^{\{\emptyset,T\}}\Gamma = \overline{\cup \int \Gamma_n d\mu}$ (we use the notation of th.35),

$$L^1_\Sigma(\{\emptyset,T\}) = \Sigma = \overline{\cup \int \Gamma_n\, d\mu}$$
$$= E^{\{\emptyset,T\}}(L^1_\Gamma) + As(\Sigma)$$
$$= \int \Gamma\, d\mu + As(\Sigma).$$

It remains to prove $\Sigma = w \int \Gamma\, d\mu$. By theorem 35

$\delta^*(x|\Sigma) = \int \delta^*(x|\Gamma(t))\ \mu(dt)$. So $\Sigma = w \int \Gamma d\mu$

2) The equality $w \int \Gamma\, d\mu = \overline{\int \Gamma\, d\mu}$ follows from the facts that

$$\delta^*(x|\int \Gamma\, d\mu) \ge \delta^*(x|\int \Gamma_n\, d\mu)$$

and $\delta^*(x|\int \Gamma_n\, d\mu) \nearrow \delta^*(x|\Sigma) = \delta^*(x|w\int \Gamma\, d\mu)$.

Thus $\overline{\int \Gamma \, d\mu} \supset w \int \Gamma \, d\mu$. The converse inclusion is obvious.

Remarks. 1) We give an example where $\int \Gamma \, d\mu$ is not closed.

Let $T = \mathbb{N}$ and μ be the measure such that $\mu(\{n\}) = 1$ for every n.

Let $F = F' = \mathbb{R}^2$ and $\Gamma(n) = \overline{co}\{(0,0), (1, \frac{1}{n+1})\}$.

We write $\sum_n$ in place of $\int d\mu$. It is easy to see that the points $(x,0)$ with $x > 0$ belong to $\overline{\Sigma \, \Gamma(n)}$ but not to $\Sigma \, \Gamma(n)$.

And $As(\overline{\Sigma \, \Gamma(n)}) = \mathbb{R}_+ \times \{0\}$.

2) One can prove corollary 39 more directly without speaking of conditional expectations. But the use of the decomposition of $(L^\infty_F)'$ seems to be necessary.

40 - In the following theorem we make clear the infimum convolution property of I_g stated in remark 36.

Theorem VIII.40 - Under the hypotheses of theorem 36 suppose that there exists a $v_0 \in L^\infty_F(\mathcal{B})$ such that $I_{g^*}(v_0) < \infty$.

a) Consider the functional J defined on $L^1_{F'}(\mathcal{B})$ by

$$J(u) = \inf\{I_f(u_1) \mid u_1 \in L^1_{F'}(\mathcal{A}),\ E^{\mathcal{B}}(u_1) = u\}$$

Then $I_g = J^{**}$.

b) If moreover I_{f^*} is finite on all $L^\infty_F(\mathcal{A})$ then $J = I_g$, and if the infimum in the definition of J is finite it is a minimum.

Proof a) By theorem VII.7. I_g and I_{g^*} are mutually polar, and so it suffices to prove $J^* = I_{g^*}$. For $v \in L^\infty_F(\mathcal{B})$ one has

$$\begin{aligned}
J^*(v) &= \sup\{< u,v > - J(u) \mid u \in L^1_{F'}(\mathcal{B})\} \\
&= \sup\{< u,v > - I_f(u_1) \mid u_1 \in L^1_{F'}(\mathcal{A}),\ E^{\mathcal{B}}(u_1) = u\} \\
&= \sup\{< E^{\mathcal{B}}(u_1),v > - I_f(u_1) \mid u_1 \in L^1_{F'}(\mathcal{A})\} \\
&= \sup\{< u_1,v > - I_f(u_1) \mid u_1 \in L^1_{F'}(\mathcal{A})\} \quad \text{(by prop, 33)} \\
&= I_{f^*}(v) \\
&= I_{g^*}(v) \quad \text{(by theorem 36)}
\end{aligned}$$

b) If moreover I_{f*} is finite on all $L^{\infty}_{F}(\mathcal{A})$, by corollary 30 it is $\tau(L^{\infty}_{F}, L^{1}_{F'})$ continuous. And so the statement results from theorem I. 22 : $I_{g*} = I_{f*} \circ \mathrm{inj}$ implies that $I_g = (I_{g*})^* = J$ and if the infimum is finite it is a minimum.

Remarks 1) One could think that in the proof of b) it would be sufficient that I_{f*} be $\tau(L^{\infty}_{F}, L^{1}_{F'})$ continuous at a point $v_0 \in L^{\infty}_{F}(\mathcal{B})$. A sufficient condition for continuity at 0 has been given in theorem V.9., but under this condition I_{f*} is finite on all the space.

2) The result in b) is a kind of exactness of the conditional infimum convolution. In the following theorem we give a stronger result of exactness.

41 - Theorem VIII.41. - Under the hypotheses of theorem 36, we suppose that there exists a closed convex weakly locally compact set C in $F' \times \mathbb{R}$ which contains no line, such that $\mathrm{epi}\, f(\omega,.) \subset C$ a.e., and we suppose that $\forall x \in F$, $f^*(.,x)$ is integrable. Let $u \in L^1_{F'}(\mathcal{B})$ such that $\int g(\omega, u(\omega))P(d\omega) < \infty$. Then

a) $\forall u_1 \in L^1_{F'}(\mathcal{A})$ such that $E^{\mathcal{B}}(u_1) = u$, $f(.,u_1(.))$ is quasi-integrable and $E^{\mathcal{B}}(f(.,u_1(.)))(\omega) \geq g(\omega,u(\omega))$ a.e.

b) there exists $u_1 \in L^1_{F'}(\mathcal{A})$ such that $E^{\mathcal{B}}(u_1) = u$ and $E^{\mathcal{B}}(f(.,u_1(.)))(\omega) = g(\omega,u(\omega))$ a.e.

Proof. 1) Let $\Gamma(\omega) = \mathrm{epi}\, f(\omega,.)$, $\Sigma(\omega) = \mathrm{epi}\, g(\omega,.)$. We apply corollary 38 which says that

$$L^1_{\Sigma}(\mathcal{B}) = E^{\mathcal{B}}(L^1_{\Gamma}(\mathcal{A})) + L^1_{As(\Sigma)}(\mathcal{B}).$$

First we show that $As(\Sigma(\omega)) = \{0\} \times [0,\infty[$ a.e. As the inclusion $\{0\} \times [0,\infty[\subset As(\Sigma(\omega))$ is obvious, we have to prove the opposite inclusion. Let (x_n) be a dense sequence in F.

By prop. I.7, $As(\Sigma(\omega)) = [dom(\delta^*(.|\Sigma(\omega)))]^\circ$. Let N be a negligible set such that $\forall n$, $\forall \omega \notin N$, $g^*(\omega,x_n) \in \mathbb{R}$ (such a N exists because $\int g^*(\omega,x_n)P(d\omega) = \int f^*(\omega,x_n)P(d\omega) < \infty$).

The formula $g^*(\omega,x_n) = \delta^*((x_n, -1)|\Sigma(\omega))$ implies that if $\omega \notin N$, $(x_n, -1)$ belongs to $dom(\delta^*(.|\Sigma(\omega)))$.

Hence $As(\Sigma(\omega)) \subset \{(x',r) | \forall n, < (x',r),(x_n,-1) > \leq 1\}$

$$= \{0\} \times [0,\infty[.$$

2) Let us show now that $L^1_\Sigma(\mathcal{B}) = E^{\mathcal{B}}(L_\Gamma(\mathcal{A}))$.

Let $u_1 \in L^1_\Gamma(\mathcal{A})$ and $u_0 \in L^1_{As(\Sigma)}(\mathcal{B}) = L^1_{\{0\}\times[0,\infty[}(\mathcal{B})$.

Then $E^{\mathcal{B}}(u_1) + u_0 = E^{\mathcal{B}}(u_1 + u_0)$ and as $u_1 + u_0$ still belongs to $L^1_\Gamma(\mathcal{A})$ the formula is proved.

3) We prove now a). First $f(.,u_1(.))$ is quasi-integrable because $f(\omega,u_1(\omega)) \geq -f^*(\omega,0)$. Let

$$B = \{\omega | E^{\mathcal{B}}(f(.,u_1(.)))(\omega) < g(\omega, u(\omega))\}.$$

Remark that $\int_B f(.,u_1(.))dP = \int_B E^{\mathcal{B}}(f(.,u_1(.)))dP < \infty$.

Then the function defined on B as

$\omega \mapsto (u_1(\omega), f(\omega,u_1(\omega)))$ is an integrable selection of Γ.

So there exists $(u_2,r) \in L^1_\Gamma(\mathcal{A})$ such that $\forall \omega \in B$

$$u_2(\omega) = u_1(\omega),\ r(\omega) = f(\omega,u_1(\omega)).$$

Then $E^{\mathcal{B}}(u_2,r)$ belongs to $L^1_\Sigma(\mathcal{A})$, that is $E^{\mathcal{B}}(r)(\omega) \geq g(\omega,E^{\mathcal{B}}(u_2)(\omega))$ a.e.

For $\omega \in B$ that means

$$E^{\mathcal{B}}(f(.,u_1(.)))(\omega) \geq g(\omega,u(\omega)) \text{ a.e.}$$

Thus B is negligible and a) is proved.

4) We prove now b). As $(u,g(.,u(.)))$ belongs to $L^1_\Sigma(\mathcal{B})$, by 2) there exists $(u_1,r) \in L^1_\Gamma(\mathcal{A})$ such that

$$E^{\mathcal{B}}(u_1,r) = (u,g(.,u(.))).$$

That means $E^{\beta}(u_1) = u$ and $E^{\beta}(r) = g(.,u(.))$. But by a)

$$E^{\beta}(r) \geq E^{\beta}(f(.,u_1(.))) \geq g(.,u(.))$$

and so $g(.,u(.)) = E^{\beta}(f(.,u_1(.)))$.

BIBLIOGRAPHY

1. BISMUT, J-M. Intégrales convexes et probabilités. J. of Math. An. and Appl. 42-3 (1973) 639-673.

2. BOURBAKI, N. Intégration Ch. I,II,III,IV 2nd édition.

3. BOURBAKI, N. Théories spectrales Ch. I, II.

4. CHOQUET, G. Lectures on Analysis. Benjamin (1969).

5. DAURES, J-P. Version multivoque du théorème de Doob. Ann. Inst. H-Poincaré IX-2(1973)167-176

6. DINCULEANU, N. Vector measures. Pergamon Press (1967).

7. DIXMIER, J. Sur certains espaces considérés par M.H. Stone. Summa Brasil Math 2 (1951) 151-182.

8. DUBOVITSKII, A. Ya. MILIUTIN, A.A. Necessary condition for a weak extremum in problems of optimal control with mixed constraints of inequality type. Zh. Vychisl. Mat. i Mat. Fys. 8(1968) 725-779.

9. DUNFORD, N. SCHWARTZ, J.T. Linear operators. Part I. John Wiley (1964).

10. FHIMA, G. Applications intégrales et intégrandes convexes. Décomposition du dual de L_E^∞. Thèse de 3ème cycle Paris (1974).

11. HIRIART-URRUTY, J-B. Etude de quelques propriétés de la fonctionnelle moyenne et de l'inf-convolution continue en analyse convexe stochastique. C.R. Ac. Sc. 280 (1975) 129-132.

12. HUFF, R.E. The Yosida-Hewitt decomposition as an ergodic theorem in Vector and Operator Valued Measures edited by Tuckerand Maynard Academic Press. 1973 p.133-139.

13. IOFFE, A.D. LEVIN, V.L. Subdifferentials of convex functions. Trudy Mosk. Mat. Ob. 26 (1972) 3-73.

14. IOFFE, A.D. TIHOMIROV, V.M. Minimisation of an integral functional. Funkt. an i priloj 3-3 (1969) 61-70.

15. IONESCU TULCEA, A. and C. Topics in the theory of lifting. Springer Verlag (1969)

16. LEVIN, V.L. Lebesgue decomposition of functionals on L_X^∞. Funkt. Anal. 8-4 (1974) 48-53.

17. LEVIN, V.L. Convex integral functionals and theory of lifting. Uspehi Mat. Nauk. XXX-2 (1975) 115-178

18. NEVEU, J. Bases mathématiques du calcul des probabilités. Masson (1964).

19. NEVEU, J. Convergence presque sûre de martingales multivoques. Annales Inst. H. Poincaré VIII-1 (1972) 1-7.

20. PALLU DE LA BARRIERE, R. Etudes de quelques espaces liés à l'intégration multivoque C.R. Acad. Sc. 278 (1974) 1491-1494.

21. PERESSINI, A.L. Ordered topological vector spaces Harper Row (1967).

22. ROCKAFELLAR, R.T. Convex integral functionals and duality. Contribution to non linear functional analysis. Academic Press (1971) p. 215-236.

23. ROCKAFELLAR, R.T. Decomposition in $L_Y^{\infty *}$ for infinite-dimensional Y. Unpublished.

24. ROCKAFELLAR, R.T. Integrals which are convex functionals II. Pacific J. Math. 39-2(1971) 439-469.

25. SCALORA, F.S. Abstract martingale convergence theorems. Pacific J. Math. 11(1961)347-374.

26. VALADIER, M. Intégration de convexes fermés, notamment d'épigraphes. RIRO R2 (1970) 57-73.

27. VALADIER, M. Une propriété du dual de $L_{E'}^\infty$. Séminaire d'Analyse Convexe. Montpellier (1972).

28. VALADIER, M. Espérance conditionnelle d'un convexe fermé aléatoire. Comptes Rendus Ac. Sc. 273 (1971) 1265-1267 (see also Séminaire d'Analyse Convexe, Montpellier, 1972 exposé n°1)

29. VAN CUTSEM, B. Eléments aléatoires à valeurs convexes compactes. Thèse Grenoble 1971.

30. VULIKH, B.Z. Introduction to the theory of partially ordered spaces Noordhoff (1967).

31. YOSIDA, K. HEWITT, E. Finitely additive measures. Trans. A.M.S. 72(1952)46-66.

32. ZAANEN, A.C. Integration 2nd édition. North Holland (1967).

SUBJECT INDEX

. 460: O. Loos, Jordan Pairs. XVI, 218 pages. 1975.

. 461: Computational Mechanics. Proceedings 1974. Edited by
. Oden. VII, 328 pages. 1975.

. 462: P. Gérardin, Construction de Séries Discrètes p-adiques.
ur les séries discrètes non ramifiées des groupes réductifs
oloyés p-adiques«. III, 180 pages. 1975.

. 463: H.-H. Kuo, Gaussian Measures in Banach Spaces. VI,
4 pages. 1975.

. 464: C. Rockland, Hypoellipticity and Eigenvalue Asymptotics.
171 pages. 1975.

. 465: Séminaire de Probabilités IX. Proceedings 1973/74. Edité
P. A. Meyer. IV, 589 pages. 1975.

. 466: Non-Commutative Harmonic Analysis. Proceedings 1974.
ted by J. Carmona, J. Dixmier and M. Vergne. VI, 231 pages. 1975.

. 467: M. R. Essén, The Cos $\pi\lambda$ Theorem. With a paper by
rister Borell. VII, 112 pages. 1975.

. 468: Dynamical Systems – Warwick 1974. Proceedings 1973/74.
ted by A. Manning. X, 405 pages. 1975.

. 469: E. Binz, Continuous Convergence on C(X). IX, 140 pages.
'5.

470: R. Bowen, Equilibrium States and the Ergodic Theory of
sov Diffeomorphisms. III, 108 pages. 1975.

471: R. S. Hamilton, Harmonic Maps of Manifolds with Boundary.
68 pages. 1975.

472: Probability-Winter School. Proceedings 1975. Edited by
Ciesielski, K. Urbanik, and W. A. Woyczyński. VI, 283 pages.
5.

473: D. Burghelea, R. Lashof, and M. Rothenberg, Groups of
omorphisms of Manifolds. (with an appendix by E. Pedersen)
156 pages. 1975.

474: Séminaire Pierre Lelong (Analyse) Année 1973/74. Edité
P. Lelong. VI, 182 pages. 1975.

475: Répartition Modulo 1. Actes du Colloque de Marseille-
niny, 4 au 7 Juin 1974. Edité par G. Rauzy. V, 258 pages. 1975.
5.

. 476: Modular Functions of One Variable IV. Proceedings 1972.
ted by B. J. Birch and W. Kuyk. V, 151 pages. 1975.

. 477: Optimization and Optimal Control. Proceedings 1974.
ted by R. Bulirsch, W. Oettli, and J. Stoer. VII, 294 pages. 1975.

478: G. Schober, Univalent Functions – Selected Topics. V,
pages. 1975.

. 479: S. D. Fisher and J. W. Jerome, Minimum Norm Extremals
Function Spaces. With Applications to Classical and Modern
alysis. VIII, 209 pages. 1975.

480: X. M. Fernique, J. P. Conze et J. Gani, Ecole d'Eté de
babilités de Saint-Flour IV-1974. Edité par P.-L. Hennequin.
293 pages. 1975.

481: M. de Guzmán, Differentiation of Integrals in R^n. XII, 226
ges. 1975.

. 482: Fonctions de Plusieurs Variables Complexes II. Séminaire
nçois Norguet 1974–1975. IX, 367 pages. 1975.

. 483: R. D. M. Accola, Riemann Surfaces, Theta Functions, and
elian Automorphisms Groups. III, 105 pages. 1975.

484: Differential Topology and Geometry. Proceedings 1974.
ted by G. P. Joubert, R. P. Moussu, and R. H. Roussarie. IX,
pages. 1975.

485: J. Diestel, Geometry of Banach Spaces – Selected Topics.
282 pages. 1975.

. 486: S. Stratila and D. Voiculescu, Representations of AF-
gebras and of the Group U (∞). IX, 169 pages. 1975.

487: H. M. Reimann und T. Rychener, Funktionen beschränkter
tlerer Oszillation. VI, 141 Seiten. 1975.

. 488: Representations of Algebras, Ottawa 1974. Proceedings
4. Edited by V. Dlab and P. Gabriel. XII, 378 pages. 1975.

Vol. 489: J. Bair and R. Fourneau, Etude Géométrique des Espaces Vectoriels. Une Introduction. VII, 185 pages. 1975.

Vol. 490: The Geometry of Metric and Linear Spaces. Proceedings 1974. Edited by L. M. Kelly. X, 244 pages. 1975.

Vol. 491: K. A. Broughan, Invariants for Real-Generated Uniform Topological and Algebraic Categories. X, 197 pages. 1975.

Vol. 492: Infinitary Logic: In Memoriam Carol Karp. Edited by D. W. Kueker. VI, 206 pages. 1975.

Vol. 493: F. W. Kamber and P. Tondeur, Foliated Bundles and Characteristic Classes. XIII, 208 pages. 1975.

Vol. 494: A Cornea and G. Licea. Order and Potential Resolvent Families of Kernels. IV, 154 pages. 1975.

Vol. 495: A. Kerber, Representations of Permutation Groups II. V, 175 pages. 1975.

Vol. 496: L. H. Hodgkin and V. P. Snaith, Topics in K-Theory. Two Independent Contributions. III, 294 pages. 1975.

Vol. 497: Analyse Harmonique sur les Groupes de Lie. Proceedings 1973–75. Edité par P. Eymard et al. VI, 710 pages. 1975.

Vol. 498: Model Theory and Algebra. A Memorial Tribute to Abraham Robinson. Edited by D. H. Saracino and V. B. Weispfenning. X, 463 pages. 1975.

Vol. 499: Logic Conference, Kiel 1974. Proceedings. Edited by G. H. Müller, A. Oberschelp, and K. Potthoff. V, 651 pages 1975.

Vol. 500: Proof Theory Symposion, Kiel 1974. Proceedings. Edited by J. Diller and G. H. Müller. VIII, 383 pages. 1975.

Vol. 501: Spline Functions, Karlsruhe 1975. Proceedings. Edited by K. Böhmer, G. Meinardus, and W. Schempp. VI, 421 pages. 1976.

Vol. 502: János Galambos, Representations of Real Numbers by Infinite Series. VI, 146 pages. 1976.

Vol. 503: Applications of Methods of Functional Analysis to Problems in Mechanics. Proceedings 1975. Edited by P. Germain and B. Nayroles. XIX, 531 pages. 1976.

Vol. 504: S. Lang and H. F. Trotter, Frobenius Distributions in GL_2-Extensions. III, 274 pages. 1976.

Vol. 505: Advances in Complex Function Theory. Proceedings 1973/74. Edited by W. E. Kirwan and L. Zalcman. VIII, 203 pages. 1976.

Vol. 506: Numerical Analysis, Dundee 1975. Proceedings. Edited by G. A. Watson. X, 201 pages. 1976.

Vol. 507: M. C. Reed, Abstract Non-Linear Wave Equations. VI, 128 pages. 1976.

Vol. 508: E. Seneta, Regularly Varying Functions. V, 112 pages. 1976.

Vol. 509: D. E. Blair, Contact Manifolds in Riemannian Geometry. VI, 146 pages. 1976.

Vol. 510: V. Poènaru, Singularités C^∞ en Présence de Symétrie. V, 174 pages. 1976.

Vol. 511: Séminaire de Probabilités X. Proceedings 1974/75. Edité par P. A. Meyer. VI, 593 pages. 1976.

Vol. 512: Spaces of Analytic Functions, Kristiansand, Norway 1975. Proceedings. Edited by O. B. Bekken, B. K. Øksendal, and A. Stray. VIII, 204 pages. 1976.

Vol. 513: R. B. Warfield, Jr. Nilpotent Groups. VIII, 115 pages. 1976.

Vol. 514: Séminaire Bourbaki vol. 1974/75. Exposés 453 – 470. IV, 276 pages. 1976.

Vol. 515: Bäcklund Transformations. Nashville, Tennessee 1974. Proceedings. Edited by R. M. Miura. VIII, 295 pages. 1976.

Vol. 516: M. L. Silverstein, Boundary Theory for Symmetric Markov Processes. XVI, 314 pages. 1976.

Vol. 517: S. Glasner, Proximal Flows. VIII, 153 pages. 1976.

Vol. 518: Séminaire de Théorie du Potentiel, Proceedings Paris 1972–1974. Edité par F. Hirsch et G. Mokobodzki. VI, 275 pages. 1976.

Vol. 519: J. Schmets, Espaces de Fonctions Continues. XII, 150 pages. 1976.

Vol. 520: R. H. Farrell, Techniques of Multivariate Calculation. X, 337 pages. 1976.

Vol. 521: G. Cherlin, Model Theoretic Algebra – Selected Topics. IV, 234 pages. 1976.

Vol. 522: C. O. Bloom and N. D. Kazarinoff, Short Wave Radiation Problems in Inhomogeneous Media: Asymptotic Solutions. V. 104 pages. 1976.

Vol. 523: S. A. Albeverio and R. J. Høegh-Krohn, Mathematical Theory of Feynman Path Integrals. IV, 139 pages. 1976.

Vol. 524: Séminaire Pierre Lelong (Analyse) Année 1974/75. Edité par P. Lelong. V, 222 pages. 1976.

Vol. 525: Structural Stability, the Theory of Catastrophes, and Applications in the Sciences. Proceedings 1975. Edited by P. Hilton. VI, 408 pages. 1976.

Vol. 526: Probability in Banach Spaces. Proceedings 1975. Edited by A. Beck. VI, 290 pages. 1976.

Vol. 527: M. Denker, Ch. Grillenberger, and K. Sigmund, Ergodic Theory on Compact Spaces. IV, 360 pages. 1976.

Vol. 528: J. E. Humphreys, Ordinary and Modular Representations of Chevalley Groups. III, 127 pages. 1976.

Vol. 529: J. Grandell, Doubly Stochastic Poisson Processes. X, 234 pages. 1976.

Vol. 530: S. S. Gelbart, Weil's Representation and the Spectrum of the Metaplectic Group. VII, 140 pages. 1976.

Vol. 531: Y.-C. Wong, The Topology of Uniform Convergence on Order-Bounded Sets. VI, 163 pages. 1976.

Vol. 532: Théorie Ergodique. Proceedings 1973/1974. Edité par J.-P. Conze and M. S. Keane. VIII, 227 pages. 1976.

Vol. 533: F. R. Cohen, T. J. Lada, and J. P. May, The Homology of Iterated Loop Spaces. IX, 490 pages. 1976.

Vol. 534: C. Preston, Random Fields. V, 200 pages. 1976.

Vol. 535: Singularités d'Applications Differentiables. Plans-sur-Bex. 1975. Edité par O. Burlet et F. Ronga. V, 253 pages. 1976.

Vol. 536: W. M. Schmidt, Equations over Finite Fields. An Elementary Approach. IX, 267 pages. 1976.

Vol. 537: Set Theory and Hierarchy Theory. Bierutowice, Poland 1975. A Memorial Tribute to Andrzej Mostowski. Edited by W. Marek, M. Srebrny and A. Zarach. XIII, 345 pages. 1976.

Vol. 538: G. Fischer, Complex Analytic Geometry. VII, 201 pages. 1976.

Vol. 539: A. Badrikian, J. F. C. Kingman et J. Kuelbs, Ecole d'Eté de Probabilités de Saint Flour V-1975. Edité par P.-L. Hennequin. IX, 314 pages. 1976.

Vol. 540: Categorical Topology, Proceedings 1975. Edited by E. Binz and H. Herrlich. XV, 719 pages. 1976.

Vol. 541: Measure Theory, Oberwolfach 1975. Proceedings. Edited by A. Bellow and D. Kölzow. XIV, 430 pages. 1976.

Vol. 542: D. A. Edwards and H. M. Hastings, Čech and Steenrod Homotopy Theories with Applications to Geometric Topology. VII, 296 pages. 1976.

Vol. 543: Nonlinear Operators and the Calculus of Variations, Bruxelles 1975. Edited by J. P. Gossez, E. J. Lami Dozo, J. Mawhin, and L. Waelbroeck, VII, 237 pages. 1976.

Vol. 544: Robert P. Langlands, On the Functional Equations Satisfied by Eisenstein Series. VII, 337 pages. 1976.

Vol. 545: Noncommutative Ring Theory. Kent State 1975. Edited by J. H. Cozzens and F. L. Sandomierski. V, 212 pages. 1976.

Vol. 546: K. Mahler, Lectures on Transcendental Numbers. Edited and Completed by B. Diviš and W. J. Le Veque. XXI, 254 pages. 1976.

Vol. 547: A. Mukherjea and N. A. Tserpes, Measures on Topological Semigroups: Convolution Products and Random Walks. V, 197 pages. 1976.

Vol. 548: D. A. Hejhal, The Selberg Trace Formula for PSL (2, ℝ). Volume I. VI, 516 pages. 1976.

Vol. 549: Brauer Groups, Evanston 1975. Proceedings. Edited by D. Zelinsky. V, 187 pages. 1976.

Vol. 550: Proceedings of the Third Japan – USSR Symposium on Probability Theory. Edited by G. Maruyama and J. V. Prokhorov. VI, 722 pages. 1976.

Vol. 551: Algebraic K-Theory, Evanston 1976. Proceedings. Edite by M. R. Stein. XI, 409 pages. 1976.

Vol. 552: C. G. Gibson, K. Wirthmüller, A. A. du Plessis ar E. J. N. Looijenga. Topological Stability of Smooth Mappings. 155 pages. 1976.

Vol. 553: M. Petrich, Categories of Algebraic Systems. Vector ar Projective Spaces, Semigroups, Rings and Lattices. VIII, 217 page 1976.

Vol. 554: J. D. H. Smith, Mal'cev Varieties. VIII, 158 pages. 197

Vol. 555: M. Ishida, The Genus Fields of Algebraic Number Field VII, 116 pages. 1976.

Vol. 556: Approximation Theory. Bonn 1976. Proceedings. Edited R. Schaback and K. Scherer. VII, 466 pages. 1976.

Vol. 557: W. Iberkleid and T. Petrie, Smooth S^1 Manifolds. I 163 pages. 1976.

Vol. 558: B. Weisfeiler, On Construction and Identification of Graph XIV, 237 pages. 1976.

Vol. 559: J.-P. Caubet, Le Mouvement Brownien Relativiste. 212 pages. 1976.

Vol. 560: Combinatorial Mathematics, IV, Proceedings 1975. Edit by L. R. A. Casse and W. D. Wallis. VII, 249 pages. 1976.

Vol. 561: Function Theoretic Methods for Partial Differential Equatior Darmstadt 1976. Proceedings. Edited by V. E. Meister, N. We and W. L. Wendland. XVIII, 520 pages. 1976.

Vol. 562: R. W. Goodman, Nilpotent Lie Groups: Structure ar Applications to Analysis. X, 210 pages. 1976.

Vol. 563: Séminaire de Théorie du Potentiel. Paris, No. 2. Proceedin 1975–1976. Edited by F. Hirsch and G. Mokobodzki. VI, 292 page 1976.

Vol. 564: Ordinary and Partial Differential Equations, Dundee 197 Proceedings. Edited by W. N. Everitt and B. D. Sleeman. XVIII, 5 pages. 1976.

Vol. 565: Turbulence and Navier Stokes Equations. Proceedin 1975. Edited by R. Temam. IX, 194 pages. 1976.

Vol. 566: Empirical Distributions and Processes. Oberwolfach 197 Proceedings. Edited by P. Gaenssler and P. Révész. VII, 146 page 1976.

Vol. 567: Séminaire Bourbaki vol. 1975/76. Exposés 471–488. 303 pages. 1977.

Vol. 568: R. E. Gaines and J. L. Mawhin, Coincidence Degree, ar Nonlinear Differential Equations. V, 262 pages. 1977.

Vol. 569: Cohomologie Etale SGA 4½. Séminaire de Géomét Algébrique du Bois-Marie. Edité par P. Deligne. V, 312 pages. 197

Vol. 570: Differential Geometrical Methods in Mathematical Physic Bonn 1975. Proceedings. Edited by K. Bleuler and A. Reetz. V 576 pages. 1977.

Vol. 571: Constructive Theory of Functions of Several Variable Oberwolfach 1976. Proceedings. Edited by W. Schempp and K. Z ler. VI, 290 pages. 1977

Vol. 572: Sparse Matrix Techniques, Copenhagen 1976. Edited V. A. Barker. V, 184 pages. 1977.

Vol. 573: Group Theory, Canberra 1975. Proceedings. Edited R. A. Bryce, J. Cossey and M. F. Newman. VII, 146 pages. 197

Vol. 574: J. Moldestad, Computations in Higher Types. IV, 2 pages. 1977.

Vol. 575: K-Theory and Operator Algebras, Athens, Georgia 197 Edited by B. B. Morrel and I. M. Singer. VI, 191 pages. 1977.

Vol. 577: J. P. May, E_∞ Ring Spaces and E_∞ Ring Spectra. 268 pages. 1977.

Vol. 579: Combinatoire et Représentation du Groupe Symétriq Strasbourg 1976. Proceedings 1976. Edité par D. Foata. IV, 3 pages. 1977.

Vol. 580: C. Castaing and M. Valadier, Convex Analysis and Mea urable Multifunctions. VIII, 278 pages. 1977.